T0291601

THE PRINCIPLES OF
FIELD DRAINAGE

THE PRINCIPLES OF
FIELD DRAINAGE

BY

H. H. NICHOLSON, M.B.E., M.A.

*Reader in Soil Science, in the
School of Agriculture, in the
University of Cambridge*

CAMBRIDGE

AT THE UNIVERSITY PRESS

1953

CAMBRIDGE
UNIVERSITY PRESS

University Printing House, Cambridge CB2 8BS, United Kingdom

Cambridge University Press is part of the University of Cambridge.

It furthers the University's mission by disseminating knowledge in the pursuit of education, learning and research at the highest international levels of excellence.

www.cambridge.org
Information on this title: www.cambridge.org/9781316603833

First edition 1942
Reprinted 1944, 1946
Second edition 1953
First paperback edition 2015

A catalogue record for this publication is available from the British Library

ISBN 978-1-316-60383-3 Paperback

CONTENTS

PREFACE

Since the inception of the Government's schemes of assistance for the various forms of field and land drainage, advertisements have frequently appeared in the daily and weekly press for persons 'skilled, experienced, or trained in the business of the cleaning and re-grading of watercourses, farm ditches, tile and mole drainage, able to estimate costs, competent to carry out surveys and prepare schemes, including plans', and so on. This is asking a good deal. There is certainly no harm in asking, even in a country where the opportunity to acquire these accomplishments in practice has been conspicuously lacking for over a generation. The object of this little book is to interest and perhaps assist those whose business it is to organise, devise, advise on and supervise works of field drainage. Its contents are compounded of the elements of soil science, of the results of some ten years' study of field-drainage problems, especially those of heavy land, and of the experience gained since the beginning of the war, by close co-operation with many War Agricultural Executive officials and farmers in dealing with drainage problems in the field. It should not be necessary to apologise for adding to the published works on this subject. Reference to the Bibliography on p. 162 will show that over a period exceeding a hundred years it is not extensive. No effort has been made in this account to deal with the technicalities of surveying, engineering, or even the drainer's art. The author's aim has been to draw attention to the fundamentals of the subject, to portray a philosophy of draining, and to deal with the factors involved and the way in which they influence events in the soil, in the belief that the sounder and more complete the appreciation of any situation is, the more effective the measures taken to deal with it can be made.

In the preparation of this book I have been greatly indebted to my colleague, Dr E. C. Childs, who has read the script and made many helpful comments; to Mr J. Norfolk, for preparing the diagrams; and to Mr C. W. Williamson, for preparing the accompanying photographs. I owe a good deal to a number of friends among the officials of War Agricultural Executive Committees and the County Organisers in the Eastern Counties, as well as to individual farmers, for affording me opportunities to study their drainage problems. Frequent and lively discussions and arguments with Mr R. G. Kendall, a farmer and drainage contractor, have helped to keep me face to face with realities. In investigational work, too, I am very conscious of all the help and encouragement which I have received from Mr W. S. Mansfield, the Director of the University Farm, and from Professor F. L. Engledow, who has persistently and consistently fought for the recognition of the importance of field-drainage work.

My thanks are due also to those periodicals and journals which have so willingly allowed me to reprint contributions made to them of recent years, as specified in the text, and to the Ministry of Agriculture for permission to make use of the figures shewing the areas approved for mole draining, in Table XI.

H. H. N.

May 1942

PREFACE TO THE SECOND EDITION

This is a new edition in the sense that some alterations and additions have been made to the text. But they are few because, the occasion for reprinting having arisen, it is felt that the book, which was written originally in the context of the national war effort, should contain some reference to more recent events and circumstances. Accordingly slight additions have been made to Chapters I and II, while the Table on pp. 150 and 151, the Appendix giving particulars of Government assistance for Drainage Work, and the Bibliography on page 160 have all been brought up to date.

H. H. N.

September 1952

LIST OF ILLUSTRATIONS

PLATES

TEXT-FIGURES

CHAPTER I

FORMER NEGLECT OF DRAINAGE WORK. MEASURES OF GOVERNMENT ASSISTANCE

Under the stimulus of the country's need and the measures taken by the Government to encourage it, drainage work of all kinds has assumed such proportions in this country that the County War Agricultural Executive Committees have sometimes been hard put to it to keep on top of the necessary preliminaries of inspection and approval of the schemes submitted to them for grants-in-aid. It happened that when, in view of the possibility of war, steps were taken to rehabilitate the soil of our countryside as a means of producing the major proportion of our food, attention was at first given only to the lime and phosphate status of the soil, and nothing was done to encourage field draining, in spite of the obvious need for it, and of its fundamental importance.

Attention had been repeatedly drawn to the steady deterioration of field drains during the past generation by many spokesmen of the agricultural industry and by its investigators. The position in 1937 was outlined in an article by the author, reprinted here from *Agricultural Progress* of that year; it should be read in the light of current events.

The Present Condition of Drainage as a Limiting Factor in Productivity[1]

That the year 1937 has been a most trying season for farming in England is reflected in all crop reports. Adverse weather and soil conditions during the first half of the year have dominated the situation, particularly in heavy-land areas. These conditions came as a climax to several years of abnormal weather conditions, three drought years being followed by two which were excessively wet. Indeed, the summer of 1936 saw heavy-land drains running vigorously in mid-July. There were no drying periods except in August, so that the land was in poor condition to face the following wet winter. A total of 4 in. of rain sufficed to saturate land which in 1934 had been able to absorb 11 in. without drains running.

[1] Paper read at Oxford, July 1937.

In light open soils, abnormal and prolonged rises in the level of the water table have been general, with the result that flooded areas have been more widespread and extensive ponds have appeared and persisted in places where standing water has not been seen within living memory.

While it would be fatuous to pretend that all the adverse effects of this wet season might have been avoided by better field drainage, there is little doubt that the drainage properties and conditions of the soil have been decisive factors in the result. The areas to suffer were those where poor drainage conditions prevail. On this class of land field drainage has been neglected for two generations and its area steadily increases.

Drainage in England and Wales. In considering the position with regard to drainage in England and Wales as a whole, and looking back over the past hundred years, there is no doubt that field drainage went ahead in a vigorous fashion during 1840–80. Following the Public Monies Drainage Act of 1847, some £9,000,000 was spent on land drainage, about half being advanced by the Government, and subsequently repaid. In 1880, before a Royal Commission on Agriculture, the drainage engineer, Bailey Denton, in his evidence, estimated that 3,000,000 acres had been underdrained during the preceding generation, but in spite of this he put the area of wet land still in need of attention at 15,000,000 acres, rather more than half the total agricultural area of the country. The former figure was based on information provided by tile manufacturers, the latter on the geological map of the country.

No field-to-field survey has ever been carried out, but there do exist pointers on the present-day position. Prior to the passing of the Land Drainage Act of 1930 it was stated (*Journal of the Ministry of Agriculture*, 1927) on official authority that there were 1,250,000 acres urgently needing drainage, due to liability to flooding by reason of defective arterial channels, and 500,000 acres capable of improvement by small drainage schemes, i.e. the cleansing of small streams and main ditches. These figures can, however, only refer to low-lying land, immediately affected by arterial drainage. The statement stressed the fact that it took no account of land capable of improvement merely by under-drainage. It is obvious that the area of such land must be substantial.

Some years ago the National Farmers' Union made an attempt to assess the position by circularising a simple questionnaire to its county branches, whose replies indicated that the total area of land capable of improvement by field drainage amounted to about 7,000,000 acres, or one-quarter of the whole. Some branches, as for instance in Beds., Herts., East Sussex, Leicester, Northants. and Worcester, put their estimates at 50–80 per cent of the total; others in Hunts., Cambs., Essex, Bucks., Middlesex, Wilts., Isle of Wight and East Yorks. at 30–50 per cent. The biggest figures were associated with heavy soils and low-lying river valleys.

Data derived from a closer examination of the problem were recorded by R. McG. Carslaw in 1931. He dealt with some 170,000 acres, comprising 1,000 individual farms, spread roughly equally over heavy land, loams, and light soils in the eastern counties. His summary is as under:

Area in need of drainage		Area drained in previous 5 years
Heavy land	26 per cent	8 per cent
Loams	13 ,,	4·5 ,,
Light land	3 ,,	0·25 ,,
Whole area	14 ,,	5 ,,

The results of a field-to-field survey appear in a report of Hunter Smith and Williams in 1932 on the Barnet and District Grassland Competition. Primarily concerned with heavy London Clay soils, carrying much poor grassland, the report is nevertheless a telling commentary on the significance of field drainage. Without recounting the findings in detail, it may be sufficient to say that the judges found that 51 per cent of 5,800 acres, in 32 holdings, suffered from defective drainage. In a similar fashion, the Herts. county authorities have arrived at figures of 5–12 per cent on the light soils in the middle of the county and 16 per cent on the Boulder Clay in the north.

Government Assistance for Drainage. One of the most significant and cheering features of the new national policy for agriculture is that at last attention is being directed to the soil itself and to the fundamental factors of fertility, as witness the amounts to be spent on lime, slag and drainage. The national contribution promised for the last-named is £140,000, described by Mr Lloyd George as 'preposterous' in the face of his estimate of 1,000,000 acres of land rotting and souring through lack of drainage. Though the contribution is small, it is to be welcomed as a recognition of the importance of drainage. The Minister rightly says that in drainage we must work from the sea up to the hill and only when the main rivers are got into good order can the Government proceed up the hill and do something to help them. This view is presumably based partly on the danger of flooding in the lower reaches and the possible increased risks consequent on improved field drainage.

It is a specious argument that the better the land is drained the easier it is for surplus water to get away and therefore the greater the risk of flooding lower down. But the problem is not so easy as this. It certainly cannot be maintained that the big floods in various quarters in the last generation were due to the improvement of field drainage in the country as a whole. Some of the worst localised floods occur in towns and urban districts where the surface of the earth has been rendered absolutely impermeable by roofs, pavements and roads, and run-off as a result is instantaneous. On a former occasion evidence has been produced of the effect of increased permeability in soils in reducing the peak of field-drain run-off and in spreading out the run-off over a longer period. There are very strong grounds for the opinion that improved field drainage will lessen the risk of dangerous floods in main channels. In any case the importance of arterial channels should not be made an argument for inaction on the land behind them. Arterial drainage can give direct benefit only to a fraction of our ill-drained land.

To return to this £140,000. This, of course, is not all that is being done. As the Minister has pointed out, the Catchment Boards, since their inception in 1930, have accomplished a great deal, and some £6,000,000 has been spent on the main rivers. The £140,000 is a recognition of the next step, to further the work of the small internal drainage boards, county councils, and rural district councils, struggling with the maintenance of subsidiary streams of less importance, but nevertheless public drains vital to agricultural land.

As an instance of this, one might quote the case of the Chesterton Rural District Council, which functions as a drainage authority around the town of Cambridge. Its area covers 112,000 acres with a rateable value of £108,000. Out of the rates the equivalent of 8d., or round about £3,000 per annum, is spent on drainage. Twenty men and a foreman are constantly employed in maintaining some 200 miles of minor streams and ditches. This burden on the locality is by no means a light one, yet it is admitted that much more ought to be done than is in actual fact. Although cases such as this are not of common occurrence, it is obvious that there is still a mass of work to be done between the main drains of the Catchment Boards and the field drains of the farmer.

Field Drains. And this is where, for the present, Government assistance ends, although we are assured that those responsible are aware of what still lies untouched higher up the hill. There still remain to be considered the field ditches and the actual land drains. Both these concern the farmer immediately, and for most farmers they are of much more vital interest than arterial drainage works. Unfortunately, for over a generation now, there has been an increasing tendency from various causes to neglect them. Wherever agricultural opinion expresses itself on the drainage question, the cleansing of ditches is invariably stressed. There can be no doubt that the neglect of ditches has led more than anything else to the loss of much excellent tile-drainage work of last century. To omit to carry out even the annual clearance of weeds will result in many outfalls being submerged the following winter when the drains begin to run. Further neglect causes accumulations of silt, the blocking of the drains and the water-logging of the land they are meant to serve.

Even to-day, thorough cleaning of ditches would restore many such old drains to efficient service again, particularly in light-land areas. The same is not necessarily true in heavy land, however, as will be seen. Instances are not uncommon in light-land areas where the removal of anything up to 3 ft. of silt and other accumulations have revealed extensive systems of tile drainage, in good order, which have functioned effectively as soon as their outfalls were uncovered.

The different conditions in heavy land may be illustrated by the case of an area of typical Gault Clay taken over by the Cambridge University Farm in 1930. The ground is all low-lying, but not flat, most fields having a moderate fall. It is all dependent on a small stream maintained by the local authority. The area was in a semi-derelict state and was immediately surveyed with a view to drainage. The ditches at that date were in poor condition, but the surveyor found 12 outfalls in a length of 900 yd., some running, some not. The whole area was then mole drained, tile mains and new outfalls were provided, and the ditches were cleaned. The same ditches were cleaned again and deepened by 12–24 in. last spring. The operation revealed 35 separate outfalls.

This particular cleansing was carried out on a total of 1,232 yd. of ditch, disclosing altogether 55 outfalls, of which only 3 belong to the present drainage system. They occurred at varying depths between 27 and 46 in., but mostly at about 36 in. The type and condition of the outfalls is indicated below:

Diameter	Total number	Silted or otherwise choked	Clear
Less than 2 in.	20	18	2
2–3 ,,	15	12	3
3–4 ,,	5	3	2
4–5 ,,	13	0	13
5–7 ,,	2	0	2

The facts of this particular case have an important bearing on the whole question of drainage of heavy land. Over what period these many drains have been accumulating is uncertain, but they obviously represent many separate efforts to solve the problem. None of them has achieved more than a temporary mitigation of the evil due to the nature of the clay itself. Clays are impermeable and it is only the surface layer of soil which is endowed with any degree of permeability, which, alas, diminishes rapidly with depth.

One of the factors producing this permeability is seasonal weather changes, and it is common experience that heavy land drains more freely after a droughty summer than after a moist one. So this natural permeability is primarily a seasonal variant. The different methods of field drainage first of all provide a graded system of channels by which the percolating water can escape, but they also produce a greatly enhanced permeability in the immediate vicinity of these channels due to the inevitable disturbance and opening of the soil which is caused, whether it be by the drainer's spade or by the passage of the mole plough. The value of the original permeability will continue to wax and wane according to the season, but the artificially induced permeability above the channels will steadily decrease.

In drainage work, the tendency is to expect drains to last a long time. Something approaching permanency has been achieved, certainly, in the case of light land, but the heavier the land the less durable has any form of drains proved to be. The same attitude is frequently encountered with respect to mole draining. To some it is a matter of pride that their mole drains last 10, 15 or 20 years.

Plate I *a*. Winter flooding. (*See p.* 2.)

Plate I *b*. Some of East Anglia's derelict clay land (arable), in 1939. (*See p.* 3.)

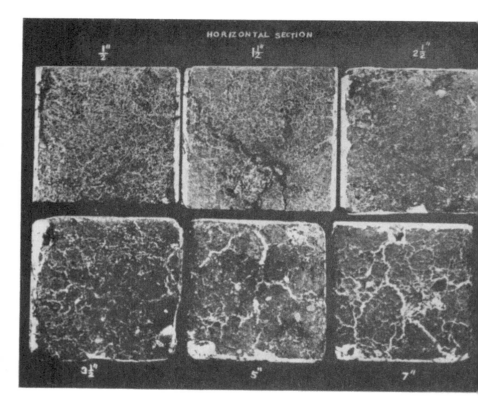

Plate II. Horizontal sections of a block of clay soil from old grassland, in its natural condition, ~~~ impregnating it and fixing it with a wax-naphthalene mixture. The sections were 6 in. ×6~~~ (*See p.* 26.)

The channels certainly may still be demonstrable, and the outfalls discharge, but the important point to consider is whether the drains are as effective as they were in the first half-dozen years of their life. If not, then the operation should be repeated as soon as circumstances allow.

Drainage of heavy land should be regarded as a cultivation rather than a permanent improvement. There are still heavy-land farmers who are prepared to indulge in steam cultivations at 25*s*. to 30*s*. per acre, and within the last 6 years there has been no lack of experimenters in gyrotilling at 30*s*. per acre. In these days it is possible to mole drain and provide semi-permanent tiled mains with sound outfalls at 40*s*. per acre. Moreover, it is possible to re-draw the moles over the same mains at 14*s*. per acre. If the operation is only effective for 5 years—and it must frequently be effective for more—the cost of the operation surely is such as to justify its wider and more frequent employment.

The position in the country as a whole to-day seems to be that the neglect of ditch cleaning has been responsible for the deterioration of field-drainage conditions in light, medium and heavy land alike. The heavier classes of land, in addition, suffer from their own peculiar drawbacks. Periodical drainage operations have been abandoned, so that the innate impermeability of heavy land has become more and more a limiting factor. Any means of tackling these two problems should be explored. At the moment, in England, the State demurs at giving assistance in drainage beyond the spheres of local drainage authorities, but in Scotland, where arterial drainage is less necessary, the farmer or landowner does receive assistance in field drainage.

It is not so many years ago that a Government scheme was operated successfully and effectively by certain county councils, for the restoration of many miles of the more important ditches. In 1929, for the mitigation of unemployment, the Ministry of Agriculture offered to contribute from 33 to 50 per cent of the cost of tile-drainage operations to landowners willing to undertake such schemes. By July 1930, expenditure of £44,000 had been approved in 421 schemes throughout the country. As things stand, even ditching is an expense of serious dimensions, and remains essentially a task for manual labour. Annual cleaning costs about 2*s*. per chain, but the thorough removal of silt and falls every 5 or 6 years may cost as much as 7*s*. 6*d*. per chain.

The practice of surface draining might well receive more attention. The use of the common plough, as and when it is needed, is a cheap and efficient way of dealing with surface water.

In East Suffolk, on many soils, it is found to be the only satisfactory method. For the field drainage of most heavy land the mole plough remains the best available means. Present-day tractor performance has reduced the cost of moling to such an extent that it can and should be regarded as a routine operation, and there are good grounds for envisaging its possibilities as a means of renovating existing drainage systems over a wider range of soil conditions than those in which it excels of itself.

Such was the picture as it appeared in 1937. Once the storm had broken, however, it was not long before the position changed, and that rapidly. One measure followed another in quick succession, until a stage was reached at which practically all drainage work qualified for a 50 per cent grant-in-aid, and money could be borrowed for the other 50 per cent on reasonable terms. On January 1st, 1940, the Minister of Agriculture announced his readiness to make grants to owners and occupiers of agricultural land in aid of the cost of approved works of mole drainage, up to 50 per cent of the actual cost of the work, including piped outlets where it was necessary to provide them, with a maximum contribution, however, of £1 per acre. It was made a condition of approval for grant, that the watercourses or ditches which were to receive the output of the mole drains should be properly cleared beforehand—indeed, before approval could be given—with the dual object of ensuring the proper functioning of the proposed mole-drain systems and of uncovering and freeing the outfalls of any former drainage systems, many of which are capable of functioning once more, either on their own or as revived by the use of the mole plough above them. Under this scheme the War Agricultural Executive Committees in the heavy arable-land counties soon got busy and mole draining became a common sight throughout the countryside.

In July 1940 further measures were promulgated to encourage and assist field-drainage work. Where ditches had got into a bad state through years of neglect it was decided to contribute 50 per cent of the cost of thoroughly re-conditioning them, but work of the kind which normally would be carried out every year, such as weed-cutting or brushing-out, was specifically excluded from the scheme. Tile draining and all other recognised methods of field draining were given assistance at the same time, the Government undertaking to pay 50 per cent of the cost up to a maximum grant of £7. 10s. per acre. The proper cleansing of the ditches or water-courses concerned was made a pre-requisite in the case of tile-drainage schemes also. Virtually, then, there was a 50 per cent grant from the Government for all drainage work. Further, mole draining and ditching were included as 'requisites' in the Agri-cultural Requisites Assistance Scheme, initiated to provide the farmer with cash resources for current expenditure. Later on, the Government announced its readiness to pay the whole cost of any field-draining work carried out, and to recover the cost, i.e. the 50 per cent falling on the individual concerned, over a period of 3 years. Even so, there existed a difficulty in the case of tenant farmers because of their limited interest in the land and the absence of any right to compensation. The Minister decided to rectify this by appropriate legislation and the Committees were thus enabled to give directions for mole draining to be carried out where they considered it necessary. Yet another helpful step was the raising of the maximum contribution to the cost of mole draining to 30s. per acre in January 1941, in order to give all possible encouragement to the provision of adequate mains to mole-drain systems, a de-sirable procedure which, however, brings the total cost per acre to some 50s. or 60s.

The urgency of the drainage problem was at last fully recognised and many obstacles in the way of dealing with it have now been removed. It must be remembered that the measures outlined above are designed to deal with conditions in individual fields. Compre-hensive measures have also been forthcoming to deal with rivers and larger watercourses, but this, the work of Catchment Boards and other public drainage authorities, has been making steady and substantial progress ever since the war of 1914–18, and is going ahead even more quickly now.

Since 1941, of course, costs have steadily, sometimes rapidly, increased. At present, 1952, the maximum grants-in-aid are £2 per acre for mole draining, £15 per acre for tile draining, and 10s. per chain for protective fencing for ditches, subject in all cases to the over-riding provision that the grant is restricted to 50 per cent of the approved cost.

CHAPTER II

THE HISTORY OF FIELD-DRAINING DEVELOPMENTS IN GREAT BRITAIN

With so much field-drainage work in progress, it is not surprising that evidence of former efforts of all kinds is daily being turned up and uncovered in the field, particularly as a result of the thorough cleaning of many old ditches. Sometimes enough is found to make the type and lay-out of the old systems fairly clear, but the date of the work is not nearly so easy to deduce. The history of field draining shews that many different methods have been employed, some of them obsolete now, but some, such as bush draining, still occasionally resorted to.

The use of covered drains in great variety reaches back some hundreds of years. Gravel-filled drains and bush draining are referred to in writings of the fifteenth century. Hollow drains of brick were in use early in the eighteenth century but tiles did not appear until the end of that century. They were not in use when Joseph Elkington in 1764 achieved fame by his discovery of the interceptor principle in draining, but began to make headway in the nineteenth century after the introduction of the plain cylindrical type of tile and of tile-making machines. There was no 50 per cent grant for tiles or tile draining in those days; on the contrary, they were subject to a tax which rose to 5s. per thousand before being abolished in 1850, largely as a result of the efforts of the Royal Agricultural Society. As late as 1843 this body was awarding Silver Medals for various forms of drain tiles, and it was probably during these pioneer days that the less common types of drain tile at present being unearthed were tried out.

It was a long time before tiles ousted the older forms of drain to any appreciable extent. Even James Smith of Deanston, 1832, ignoring the interceptor principle in favour of 'thorough' or intensive draining, used stone-filled channels in preference to pipes. His system was to run his drains parallel and close together, at 10–20 ft. intervals and 30–36 in. depth, down the slopes into main channels which were sited in the hollows or valleys, so as to constitute a tributary system covering the whole field, regardless of whether it was all wet or not. Josiah Parkes, on the other hand,

1846, with his deeper and more widely spaced drains, 48 in. deep and 20–50 ft. apart, advocated the use of 1 in. tiles. In this same year there was passed a Land Drainage Act, which led to a tremendous expansion of field-draining work throughout the country. Between 1846 and 1856, according to the eminent drainage engineer, J. Bailey Denton, some £1,000,000 was advanced by the State for field-drainage work, and subsequently was recovered in full. By the 'eighties, at least 3,000,000 acres had been tile drained by private enterprise or with the help of public funds, i.e. loans. Some of this work was characterised by blind adherence to the tenets of one or other of the chief protagonists of different systems of draining. In particular, some so-called Government draining, complying with the official requirement of some such depth as 4 ft., led to wasteful and ineffective work and gained an ill repute which has persisted in some localities to the present day.

But while all these developments had been taking place in tile draining, and indeed long before tiles had even made their appearance, the problem of the drainage of clay land had been handled along other lines. Probably the earliest description of mole draining, the operation peculiarly adapted to clay subsoils, is that of Stephen Switzer in *The Practical Fruit Gardener*, 1724. Plug draining, as it was called, produced a result similar to that of mole draining as we know it. A narrow trench was dug, a plug or string of plugs was laid in the bottom, clay was tamped in over it to build a hollow channel, and the plugs were drawn along the trench as the channel was completed. Holes were then pierced through the clay roof at intervals to admit the water, and were protected by fragments of bush or sticks. Both the production and the persistence of the channel depended on the plasticity of the subsoil clay, and the water was admitted from above via the disturbed ground of the drain trench. The logical development of a mole plough, or machine to produce these channels without digging a drain trench, followed in due course, and applications for patents are recorded at the end of the eighteenth century. These machines were drawn by large teams of horses direct, or by means of a windlass and cable, and were common in and about the county of Essex. A method of draining heavy land, intermediate between mole draining and tile draining, is well described by the Rev. Coppinger Hill in the *Journal of the Royal Agricultural Society of England* in 1843, as being a tenant's routine operation in Suffolk, usually performed in the fallow year. The minors were formed diagonally across the slope at 16 ft. intervals and 26–30 in. in depth. First of all, a strip 18 in. by 5 in. was ploughed out, followed by another 10 in. by 5 in. out of the bottom of the first; a further 9 in. single spit was removed by an ordinary type of spade and finally another 11 in. by means of a special narrow spade giving a channel 1½ in. wide at the bottom. A suitable form of packing, such as straw or fine bushes, was pushed 3 in. into the narrow trench and trodden down; loose earth was shovelled in to cover it, and the remainder was ploughed back. Such drains functioned well for 12–20 years.

In 1850 John Fowler produced a mole plough, to be drawn by means of a capstan, designed to produce a channel in the subsoil and to draw into position in its wake a string of drain tiles. Its real value, however, became apparent when it was developed for use as a mole drainer with his famous double-engined sets of tackle in 1859. This was the type of tackle used for the highly successful and standard practice of 'steam draining' in the eastern counties through the latter half of the nineteenth and the early years of the twentieth centuries, probably one of the most valuable adjuncts to clay-land farming which has ever been produced.

After the opening of the twentieth century, attention in this country began to be focused on the condition of the rivers, streams and watercourses generally, while, unfortunately, field drains and field ditches were increasingly neglected. During the war of 1914–18, and under the stimulus of the unrestricted submarine warfare on the part of Germany in 1917, matters mended for a while; much good and useful work was put in on the ditches and streams, and drain pipes began to be considered as perhaps of a similar order of importance to that of shells. This effort was short-lived; the end of the war and of the post-war boom brought about a marked diminution in the amount of field draining carried out. Progress has been made, however, in other directions. The passing of the Land Drainage Act of 1930 and the establishment of the Catchment Boards has done much for arterial drainage and has brought about steady improvement in the rivers and main streams. The object of the Act was, indeed, to achieve this end by reorganising and co-ordinating the individual efforts of some hundreds of individual drainage authorities, to check the decay of these organisations and so of the work for which they were responsible.

The post-war agricultural depression settled heavily on the clay-land farms, so that their faithful henchman, steam tackle, fell on evil days, and one set after another was laid up for lack of work. Faced with the prohibitive expense of tile draining, and unable even to meet the comparatively heavy labour charges of steam tackle, it was not long before some agriculturists began to explore the possibilities of mole draining by direct haulage of the mole plough by means of the now rapidly developing motor tractors. The Ministry of Agriculture, between 1924 and 1932, sponsored some fifty-three Mole Draining Demonstrations up and down the country in England and Wales. A number of different ploughs were evolved and improved during this time, in conjunction with various types of tractor, and much useful pioneering work was done and technical experience gained. These demonstrations were worked in conjunction with the Institute for Research in Agricultural Engineering and the County Agricultural Education Authorities. The demonstrations as such were highly successful and were witnessed by large numbers of farmers. It cannot be said, however, that they brought about any great increase in the amount of draining work carried on; nor did they result in the mole plough becoming a farmer's implement, as was hoped. Mole draining remained a contractor's job and, in the clay areas, so far as the operation was still in vogue, it continued to be performed mainly by steam tackle, until 1938. Since the inception of the land-fertility campaign, and especially since the payment of 50 per cent grants for mole draining, however, the wisdom of these early demonstrations, and the experience gained in them, is being amply repaid, and thousands of acres are being reconditioned with the help of medium-weight mole ploughs drawn by the heavier types of track-laying tractors.

The extent of the works of field drainage carried out in England and Wales since 1939 can be gauged by the fact that schemes approved for grant-in-aid up to 30 September 1950 covered, for ditching 5,968,745 acres, for tile draining 476,043 acres, and for mole draining 550,956 acres.

CHAPTER III

THE MOISTURE PROPERTIES OF SOIL.
THE INCIDENCE OF DRAINAGE

An intelligent understanding of the problems of field drainage needs more than a knowledge of surveying and engineering. The soil at any point is an entity with a strong individual character, of which the study of the physical aspects still has scope for considerable development. The moisture properties of soil and particularly its retentive or holding capacity for moisture are realised by farmers more than by those interested in flood control and water supply. The latter are conscious enough of rainfall, evaporation, run-off and percolation, but they tend to ignore the water-holding power of the soil and its significance. The Americans summarise the matter thus:

$$\text{Rainfall} = \text{Fly-off} + \text{Cut-off} + \text{Run-off.}$$

Of the rainfall at any point, some evaporates or is transpired by plants, some is held by the soil, and some percolates or drains away. The part which is held by the soil, i.e. its moisture content, is of great importance, especially in its 'buffering' effects on evaporation and percolation.

The main characteristics of the English climate are well known. The annual rainfall is, roughly speaking, lowest in the south and east, and highest in the north and west. It is spread more or less evenly through the year, and there is no very pronounced dry or wet season. Moreover, inordinately high individual falls of rain are infrequent and isolated; 1 in. in 24 hours is an uncommon event. But, while our rainfall does not shew strong seasonal variations, the humidity of the air does. The drying power of the atmosphere can be assessed in terms of the actual amount of evaporation from a free-water surface in the open. In eastern England, as a rule, this exceeds the actual rainfall each month from March to August; in the west this condition prevails for a shorter time, e.g. on the average from April to July. The position is well illustrated in Fig. 1.

It should be pointed out, however, that actual evaporation from a soil surface may not always equal that from a free-water surface, though it does so when the soil is water-logged throughout and remains so. In other conditions evaporation may be appreciably less.

The drying effect is at a maximum in July, as a rule, and its importance in connection with drainage events is very great, acting as it does through the moisture content and moisture-holding capacity of the soil. From March the soil is drying out and presently drain flows cease or become negligible (except where

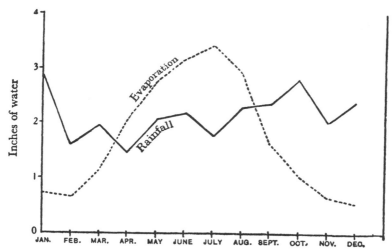

Fig. 1. Seasonal variations in rainfall and evaporation from a free-water surface in English conditions. (Greaves, 14 years' average.)

they deal with springs). The capacity of the soil to absorb and retain future rainfall increases, often to an amazing extent, and in the autumn, when rainfall once more exceeds the potential evaporation, it is some time before the soil becomes charged up to capacity and is in any need of draining. The date on which the drains begin to function again depends partly on the drying effect of the summer months and partly on the amount of rain falling in the autumn, particularly during the month of September.

To get an accurate idea of the moisture condition of soil in the field, it is necessary first of all to know something of the make-up of soil as it exists *in situ*, as distinct from its characteristics on the bench. Scientific management of the soil has suffered for far too long the handicap arising from being considered and treated as a uniform mass of inert material of definite chemical and physical constitution. This it most definitely is not, neither from one point in a field to another, nor from surface to subsoil, except in rare instances. Exposed in vertical section, the difference between surface and subsoil is obvious, and critical examination will shew

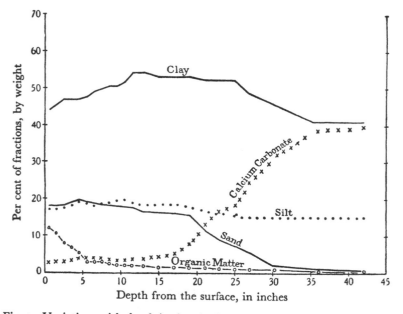

Fig. 2. Variations with depth in the physical make-up of soil in the field, as shewn by the analysis of a heavy Gault Clay soil.

more gradual but equally important changes with increasing depth from the surface. In short, a soil shews strong profile characteristics in respect of all its individual constituents and properties. In English farming conditions a soil is nearest to its natural condition after a long spell in permanent grass. Fig. 2 portrays a heavy clay in such a condition, with its main constituents set out graphically. The most important of these items in moisture relationships are the amounts of organic matter, clay, silt, sand and calcium carbonate respectively. It will be noticed in particular how the organic matter rapidly decreases with depth from the surface, and becomes very small at 6 in. The clay increases in amount to a depth of 12 in. and then, in this instance, diminishes from 26 in. downwards concurrently with a marked increase in the amount of the chalk. The amount of sand in the soil steadily falls with depth, becoming negligible at 3 ft. The effect of ploughing such a soil is to render it, to the depth of the cultivation, homogeneous in respect of all these items. There is, however, one characteristic, not easily measured, which also varies with depth and is of great importance in drainage matters. This is its state of aggregation, or tilth and structure, depending on the size of the individual soil aggregates and the way in which they are packed together. These properties decide the permeability of the soil in the mass, because while this depends primarily on the spaces between the individual soil particles, it is also greatly influenced and enhanced by the spaces between the soil crumbs and aggregates. It will thus be seen that the moisture properties of a soil can vary considerably with depth, even within that of ordinary field-drainage systems.

It is well known that a soil can hold moisture in considerable quantities even when free draining. The only signs of its presence are the darkness of the colour of the soil and a damp feeling in it. The amount of moisture that any soil can hold when properly drained is a function of the percentages of clay and organic matter in it, and particularly of the latter. This is illustrated in Fig. 3, for the heavy grassland already described, in its drained condition. The effect on it of arable farming, in making the surface soil uniform, is also evident. Light soils shew similar characteristics, but the moisture contents are substantially less throughout the profile.

If the seasonal changes in the moisture profile of a soil are studied, an idea of the importance of the effect of summer in drying out the soil can readily be obtained (see Fig. 3 *a*). The soil dries out gradually from the surface downwards except in so far as the process is checked by the intervention of rain. This drying-out proceeds to some considerable depth (see Fig. 4, where successive stages of the process in the summer of 1932 are shewn).

A perusal of the drying-out curves for 1932 shews that in June 2·8 in. were evaporated in 21 days, 1·2 in. in the next 5 days, and another inch in the next 19 days, in spite of the accession of 1·54 in. of rain. The method of arriving at these figures is crude and subject to inaccuracies, but it gives an approximate picture of events. The

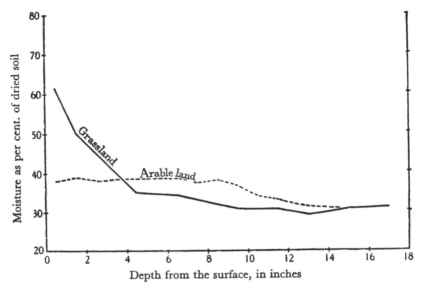

Fig. 3. Moisture content of drained clay soil in its field condition in old grassland and under arable cultivation.

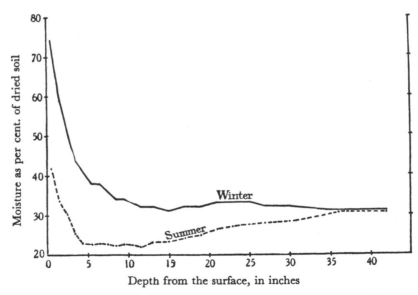

Fig. 3a. Moisture content of clay soil (old grassland) in its field condition in winter and during a summer drought.

rate of evaporation in midsummer is so high that falls of rain exert at the most only a temporary moistening of the soil. The loss by evaporation falls most heavily on the surface inches, and rain is held there, only to be evaporated again very quickly.

In the autumn, when rainfall once more exceeds evaporation, the soil is gradually re-moistened from the surface downwards.

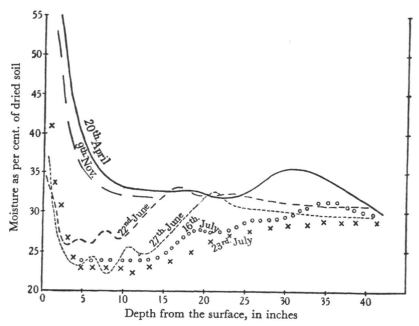

Fig. 4. Successive stages in the drying-out of a heavy clay grassland soil during a summer drought, 1932.

With the advance of winter, the daily evaporation rapidly diminishes in amount, so that each successive rainfall becomes relatively more potent in penetrating and moistening the soil. This is illustrated in Fig. 5, which shews the re-moistening of a heavy grassland soil after the dry summer of 1934. It will be observed that the drying-out process was perceptible to a depth of 70–80 in. by August 4th. Between then and October 1st, 3·76 in. of rain fell without markedly wetting the soil. By November 16th, the soil was fairly moist at the surface, and the moisture front had penetrated to a depth of 20 in., 3·1 in. of rain having fallen meanwhile.

The negligible effect of the rain of August and September is accounted for by the continuation of evaporation during that period. By December 7th, a further 1·33 in. of rain had carried the re-moistening of the soil to a depth of 26 in.; by December 21st, after 0·81 in. of rain, the moisture front was at 40 in., and 2·07 in.

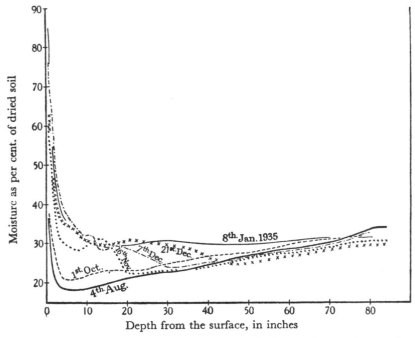

Fig. 5. Successive stages in the re-moistening of a heavy clay grassland soil—following its drying-out during a summer drought—under the influence of autumn and winter rainfall, 1934–35.

between then and January 8th, 1935, restored the moisture content of the whole profile to its normal winter condition. The drains on this land began to run on December 28th. It is clear that the summer had created a reservoir effect in the soil whereby it could absorb so much rain, that 11 in. fell during the next 5 months without there being any surplus to run away via the drains. Even without taking into account the considerable evaporation of the autumn months, it is found that an ordinary summer will create in a heavy soil a deficiency of moisture, equivalent to 4–6 in. of rain.

The moisture-holding capacity of soils, even when freely drained, is thus seen to be considerable. In the case of the heavy soils described above, examination of the moisture profiles shews that 60 in. of the soil *in situ* contains in its moist winter condition, even after draining, the equivalent of 30–35 in. of rain.

The difference between the normal winter moisture content of the soil and that at any date in the process of re-moistening represents the amount of water which the soil can absorb before drainage or percolation to the water table can begin (except in the unlikely event of the rate of rainfall exceeding that of absorption). This amount will be determined first of all by the type of soil, then by the nature of the summer (its rainfall and humidity). Light soils have a smaller reservoir capacity and the deficiency is made up more quickly than in the case of heavier soils under the same rainfall. Other factors which influence soil moisture content to a marked degree at some times of the year are the type of cropping, the yield, and the nature and incidence of cultivations during the preceding months. A good cover of vegetation always serves as a mulch and tends to keep the soil below it more moist, in spite of the amount removed by the transpiration of the crop itself. Surface cultivations, e.g. fallowing, establish a mulch of soil, although in this instance it is in the form of big clods, which conserves the moisture in the subsoil in a striking fashion; while the consolidation of a tilth to the surface brings the loss by evaporation to a maximum. These effects are reflected ultimately in the date at which the drains begin to function in the following winter. It will be seen then that it is possible to forecast, at all events in terms of rainfall, the incidence of drainage run-off in any particular season for individual fields.

The way in which one year differs from another, as regards drainage events, is well shewn in Table I, giving the monthly and annual totals of rainfall for the years 1932–40 at Cambridge in conjunction with the dates on which the heavy- and light-land drains respectively began to run.

Run-off commenced at dates varying from September 20th in a wet year (1937) to April 27th after a long drought (1933). This is the general trend of events, but is not the only factor concerned. The effect of the preceding summer rainfall depends very much on its incidence relative to the humidity and drying spells of the same period, the rainfall and humidity of the autumn, especially the months of September and October, and the amount of individual falls of rain during that period, as shewn in Table II.

Low rainfall in summer, June to August, generally results in the commencement of drain activity being later, especially if it is followed by low autumnal rainfalls, e.g. in 1933, 1934 and 1940.

TABLE I. *Shewing monthly rainfalls and date of commencement of drain flows at Cambridge*

Year	Jan.	Feb.	Mar.	Apr.	May	June	July	Aug.	Sept.	Oct.	Nov.	Dec.	Total	Heavy land	Light land
1932	0·8	0·2	1·3	2·4	3·8	0·8	2·2	2·5	1·7	3·7	1·5	0·4	21·3	Oct. 11	
1933	1·1	1·3	2·1	1·4	1·8	1·8	1·7	0·9	1·7	1·6	1·3	0·4	17·1	Apr. 27	Not observed
1934	0·8	0·4	1·1	2·2	0·7	1·1	1·5	2·1	1·9	1·6	1·5	3·5	18·4	Dec. 28	
1935	1·9	1·6	0·3	2·7	0·7	2·1	1·0	1·7	4·8	2·2	3·5	2·8	25·3	Sept. 30	Sept. 30
1936	2·9	2·1	0·7	1·3	0·9	3·3	5·3	0·5	2·4	1·8	2·4	1·7	25·3	Nov. 3	Sept. 26
1937	2·9	3·1	2·7	2·5	3·1	1·6	2·5	0·7	3·0	2·2	1·8	2·3	28·4	Sept. 20	Sept. 20
1938	2·5	0·4	0·3	0·2	1·9	0·6	1·2	1·2	3·7	2·2	1·8	2·7	18·7	Dec. 17	Oct. 13
1939	4·3	0·3	2·8	2·9	1·0	1·9	1·3	2·0	2·1	4·0	2·8	1·1	26·5	Oct. 30	Oct. 16
1940	1·7	2·4	2·8	1·2	1·4	0·6	2·9	0·2	1·2	1·5	6·2	1·5	23·6	Nov. 15	Nov. 15

Date of first drain flow

TABLE II. *Shewing the distribution of the rainfall relative to the commencement of drain flow*

Year	Rainfall June-Aug.	Rainfall Sept.	Rainfall Oct.	Commencement of drain flow	Preceding rainfall from mid-Aug.	Preceding rainfall from Sept. 1st
1932	5·5	1·7	3·7	Oct. 11	4·82	2·73
1933	4·4	1·7	1·6	Apr. 27	10·05	9·45
1934	4·7	1·9	1·6	Dec. 28	9·41	8·32
1935	4·8	4·8	2·2	Sept. 30	6·40	4·81
1936	9·1	2·4	1·8	Nov. 3	4·50	4·44
1937	4·8	3·0	2·2	Sept. 20	3·04	3·00
1938	4·0	3·7	2·2	Dec. 17	8·51	8·05
1939	5·2	2·1	4·0	Oct. 30	6·41	5·95
1940	3·7	1·2	1·5	Nov. 15	8·06	7·97

The autumn rainfall is the more potent factor, however, and even a wet summer, e.g. 1936, need not necessarily cause early run-off, if followed by low rainfalls in autumn. But at any moment a heavy downpour in the autumn may completely neutralise the reservoir effect created by a dry summer and set the drains running, e.g. 1935, with 4·8 in. in September, and 1940 with 6·2 in. in November.

It will be noticed, too, in Table I, that during the years 1935–40, the light-land drains on three occasions began to run several weeks earlier than those in the heavy land, and never later. Bailey Denton drew attention to this same point in 1861 as a normal occurrence, his own records at Hinxworth shewing a difference of 2 months in favour of the open soils. The smaller amounts of clay and organic matter in such soils is accompanied by a smaller moisture-holding capacity, so that the deficiency created by summer drying can soon be wiped out by smaller rainfalls than in the case of clay soils. The net result is that under similar rainfalls, i.e. in the same immediate locality and where the drains are not served by perennial springs, those of the light land often begin to run much earlier than those of the clay.

CHAPTER IV

PERCOLATION. PERMEABILITY. THE WATER TABLE. THE SPECIAL CASE OF CLAY LAND

In the autumn or winter, once the moisture content of a soil has reached its maximum for the freely drained condition, any fall of rain will produce a surplus in the soil. This is visible to the eye, and it moves or tends to move downwards through the body of soil, under the influence of gravity, as percolating water. The freedom with which it can do this depends on the physical make-up of the soil. The deciding factor is the calibre of the pore space, or the size of the spaces between the individual soil particles. In the simplest cases these depend on the sizes of the particles themselves. General terms such as gravel, sand and silt imply size differences; the conventional definitions in use by soil workers are:

Gravel	More than 2 millimetres in diameter
Coarse sand	2 to 0·2 ,, ,,
Fine sand	0·2 to 0·02 ,, ,,
Silt	0·02 to 0·002 ,, ,,
Clay	Less than 0·002 ,, ,,

In its natural condition, a geological deposit or horizon, i.e. the material underlying the soil and subsoil, may consist largely of one of these grades of particles, e.g. coarse, fine, or mixed gravels, sands of various grades, or clay. Soils and subsoils, however, are characterised by less homogeneity in their constitution; they are generally blends of all the different-sized particles, though one of these is often so prominent that it dominates the properties of the whole mass. Of all the mineral particles found in soils, clay has the most powerful influence; the percentage of clay present largely decides the physical properties of the whole mass, though, as will be seen, its influence may be somewhat tempered by the amounts of chalk and organic matter also present. The physical make-up of a soil material can be expressed by its physical or 'mechanical' analysis in detail, or it may be summarised qualitatively in terms indicating its texture. No conventional definition of these texture terms has been adopted in this country, but Table III gives an idea of the meaning of some of them.

Clay particles are extremely minute, so that even when dry and rigid, like grains of sand, the calibre of the pore spaces is so small that the rate of percolation of water through them is negligible. They are, however, colloidal; and when they absorb moisture, as has been pointed out, the mass swells enormously and becomes tightly packed, plastic and impermeable. The higher the amount of clay in a soil or subsoil, the less permeable it is likely to be.

TABLE III. *Shewing the composition of some common soil types in percentages by weight*

	Coarse sand	Fine sand	Silt	Clay	Calcium carbonate	Organic matter
Coarse loose sand	71	17	4	2	0	2·5
Loamy sand	45	33	8	8	0·2	3·0
Light loam	36	32	11	15	0·1	3·2
Medium loam	24	33	10	23	0·5	5·0
Heavy loam	25	22	14	31	1·2	4·0
Clay	18	19	14	41	0·5	4·7
Heavy clay	6	11	17	55	1·6	4·5
Chalky soil	8	17	4	9	58·0	2·0
Peaty soil	1	1	3	17	0·6	64·0

Through a bed of gravel, water will pass rapidly. Through similar beds of coarse sand, fine sand or silt the rate of percolation is successively slower, while through a bed of clay, partly owing to the smallness of the spaces between the particles and partly owing to their own peculiar properties, it is nil. It will be seen that permeability is in itself a very variable characteristic, and drainage problems can arise through differences in permeability which in themselves are not very great.

The examples just quoted are concerned with differences of texture, but in the realm of soils there are others, equally important: those of tilth and structure. All soils other than loose sands, owing to the clay and organic matter in them, have the power, under the influence of weathering, to form compound particles, crumbs or structural units. The weathering influences referred to comprise alternations of wet and dry, of frost and thaw; aeration and leaching; the root action of plants; and the effects of earthworms and all other soil organisms. Their combined effect is at a maximum at the surface and tails off with depth; thus producing physical differences throughout the soil profile. These physical aspects of the soil are of great importance in drainage, and can be seen most prominently in clay soils, in their natural state or under permanent grass. The tilth layer is highly permeable, but the subsoil is the reverse. The surface layer or topsoil consists of crumbs with a goodly proportion of pore space, giving an open or permeable structure. The size of the crumbs or units and their arrangement changes with depth, and at a few inches from the surface, crumbs give place to a structure of clods, of large units fitting close together with very little interspace through which water can percolate. Such spaces do exist, however, between the individual units; they can be seen when a lump of such subsoil is broken by hand, the surfaces of the units shewing up smooth and glazed by the presence of moisture. As the soil dries out in summer, the contraction of the clay causes these planes of division to become visible as fissures, a characteristic of heavy land in drought. The more marked and prolonged the drought the deeper and more numerous the fissures become, i.e. the clay acquires an increased permeability which is realised during the following winter in a better drained condition of the land. Something of the nature of this tilth and structure of a soil can be seen in Plate II, obtained by cutting and drying out blocks of the soil in its natural state, impregnating them with a hot wax-naphthalene mixture, cooling until set, then cutting and polishing the cross-sections thus obtained. It must be borne in mind, however, that this is the condition in the dried-out and not the usual moist state.

Percolating water, in its descent through the ground, may encounter an impermeable layer within a few inches of the surface as in clay land, or within a foot or two in the form of chemical hard-pan, or within a few feet as a different geological formation. On the other hand, if the formation is permeable and of great thickness, the water will move steadily downwards to considerable depths without hindrance and will rarely become the cause of a drainage problem to the soil above it. Such, for instance, is the case in much of our chalk country—the farmer's chief problem being to retain a sufficiency of moisture in his soil to grow a crop, and not to get rid of a surplus. But if the percolating water encounters an impervious stratum, it begins to pile up on it, filling the pore space of the overlying deposit as a rising water table. If the surface of this impervious layer is not level, then the water will begin to move along the greatest slope of that surface. Perhaps the simplest case is that of a bed of gravel in an elevated situation lying on a clay stratum, as shewn in Fig. 6. Excess of rain percolates freely downwards until it arrives at the clay, when it piles up in the gravel above and also begins to move outwards, partly under the influence of gravity and partly under hydrostatic pressure, along the sloping surface of the clay, to appear ultimately at the edges of the gravel deposit causing wet marshy ground or a line of springs. In the middle of the gravel area the water table may never come near enough to the surface to be a nuisance or to necessitate field drains, but round the edges there is bound to be a belt of land in such plight, broader in some years than in others, according to conditions of drought and rainfall as already explained.

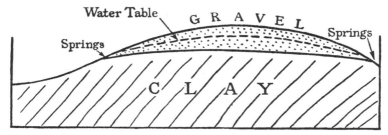

Fig. 6. Cross-section of a gravel deposit on clay, with its water table and the position of its springs.

Similar circumstances occur wherever the outcrop of a permeable formation lies above that of an impermeable one, a frequent occurrence in the south-eastern half of England (see Fig. 7). In studying this diagram it is important to bear in mind the dimensions as they occur in real life. The outcrops of the individual strata may be from 1 to 6 miles broad, the thickness of the strata from 50 to 500 or more feet and the difference in altitude between one boundary and the next much less, so that it will be seen that the water table may come near to the surface in a narrow belt within the confines of a single field or it may produce similar drainage conditions over a wide belt of country, depending on the surface topography.

It will be appreciated, too, that in times of excessive rains even a small difference in permeability between two adjacent strata may cause a temporary hold-up of water, that is to say, a rising water table, with consequent drainage trouble in the soil.

The fluctuations in the level of the water table in some undrained but permeable soil are illustrated in Fig. 8, in conjunction with the daily rainfalls responsible for them. In Fig. 9 are shewn the fluctuations of the level at one point during a number of successive years, the rise taking place at widely differing dates attributable largely to the dryness or otherwise of the preceding summer and the manner of the incidence of the rain in the autumn months. It will be noted that in the winter 1934–35 the water never appeared in this particular test well. As a rule the water table asserts itself

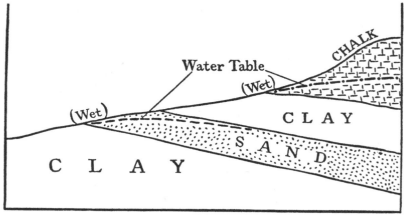

Fig. 7. Cross-section of alternating strata of permeable and impermeable deposits, with the position of the water tables, spring lines and wet land.

markedly between mid-January and early March and its connection with February 'fill-dyke' needs little explanation. The similarity in the behaviour of the water table at two points quite independent of each other under the same rainfall is well illustrated in Fig. 10; in one case the test well is in gravel lying on clay in an elevated position; in the other it is in gravel lying on a marl within 100 yd. of a river and not more than 1 ft. higher than the river bank, at a place 7 miles away from the first.

In open soils and subsoils there is a gradual rise of the water table in autumn and winter and a steady fall in spring and summer, easily demonstrable in test wells. But in clay land, the sequence of events appears different. There is little sign of excessive water in the soil until one becomes aware of it lying on the surface in pools.

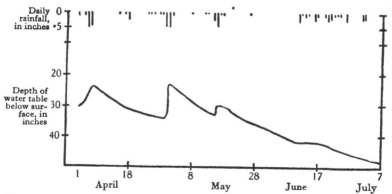

Fig. 8. Fluctuations of the water table in a gravel formation lying on clay, and the rainfalls causing them, 1939.

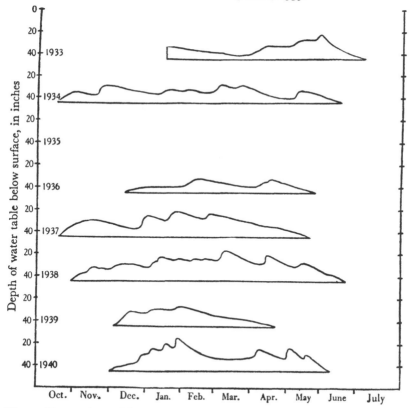

Fig. 9. Fluctuations of the water table at one point over a succession of years.

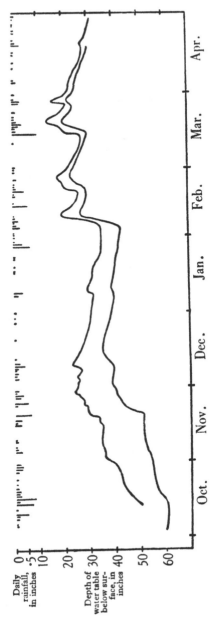

Fig. 10. Fluctuations of the water table at two points, 7 miles apart, together with the rainfalls causing them, 1939–40.

The water table suddenly appears at the surface without warning. The explanation lies in the fact that only the topsoil in clay land is permeable and percolation of free water, practically speaking, can only proceed to the base of the tilth layer. Here it encounters the swollen, homogeneous, impervious, subsoil clay, through which the only means of passage are occasional worm holes, root channels, or the vestiges of the summer fissures. The result of this is that a superficial water table appears in the topsoil and makes itself visible above the surface in all depressions and low places, where it is augmented by seepage from the surrounding higher parts.

The drainage properties of heavy soils were described by the writer in the *Transactions of the Third International Congress of Soil Science*, Vol. I, 1935, as follows:

THE DRAINAGE PROPERTIES OF HEAVY SOILS

The most profitable method of attack on any drainage problem in the field is to examine the soil profile and the underlying parent material. Light soils, generally speaking, are endowed with fair permeability throughout the profile, a permeability capable of precise definition, leading to possibilities in the way of mathematical treatment of the problem, whose solution lies in the successful regulation of the level of the water table. In heavy soils profile examination reveals structural features calculated to counteract disabilities due to texture. The structure of the soil suggests the influence of three factors in particular, seasonal weathering, biological activities and tillages. The existence and mutability of the surface tilth or crumb condition, the rapid changes in structure with increasing depth, and the marked contrast between soil and subsoil, are alike apparent. Permeability obviously decreases rapidly with depth, while the ordinary manifestations of the water table and its fluctuations are absent.

The field-moisture régime of heavy soils is a feature which merits closer investigation. Here again strong profile characteristics appear. On a heavy clay soil at Cambridge under an annual rainfall of 22 in. in normal years, without marked seasonal incidence, the winter condition of this soil is remarkably steady, once the autumn rains have moistened it. In grassland, the moisture content decreases rapidly from the surface to a depth of 5 in., then more slowly to 10 in., below which it is fairly constant. Fluctuations due to individual falls of rain during the winter are most marked at the surface, and are confined to the top 5 in. There is no substantial difference between the moisture profiles of drained and undrained ground in the case of this heavy soil except in the surface inches, during and immediately after falls of rain. In arable-land profiles, the gradient in the moisture content of the surface soil, so prominent a feature in grassland, is absent. Each summer, the drying-out of the soil proceeds from the surface downwards and, as may be expected, to varying depths. In 1934 this was measurable to 80 in. The autumn re-wetting proceeds similarly and the drains do not run until it has been completed.

Free water is never encountered in this soil except in the surface inches, during the actual fall of the rain, or in hollows where water is to be seen lying on the surface. Even in such conditions one has to penetrate only a few inches to reach horizons devoid of percolating water.

Everything points to the tilth layer as being the only horizon of the soil in which effective percolation can take place. Slater and Byers called attention to the function of cleavage planes in the soil structure in this connection, but it appears probable that these may be of little significance compared with the tilth layer. Attention has been called to the importance of the structure of the soil and it is on this that permeability probably depends. Investigations of both characteristics support this. Mistchenko's method of fixing soils has shewn us that in heavy clay grassland there is a strongly developed crumb structure at the surface, which fades out rapidly with depth and is replaced by comparatively infrequent cleavage planes. The necessity of drying-out the soil blocks before impregnating them, with the inevitable shrinkage which results, causes the channels below the tilth layer, although infrequent, to appear unduly prominent. As an indication of the dimensions of these channels, the method fails to give satisfaction, but there can be no doubt as to the significance of their relative frequency at the various depths.

These studies of the structure of heavy soils shew up strikingly the open nature of the tilth in arable land and the sudden change, at the depth of the tillage, into an impermeable subsoil. In arable land, structural conditions in the soil, the result chiefly of tillages, are very different from those in grassland. A definite and more or less uniform crumb is produced to a certain depth, at which there is a sudden transition to a markedly impermeable subsoil. This being the case, one would expect the drainage properties of the soil, in this condition, to approach somewhat those of a light open soil. Comparable outfall records shew that from light soils the flow is more steady and continuous throughout a drainage season, while from heavy soils it is characterised by marked and rapid flushes following each downpour, with negligible flow between whiles. The outfall records for arable and grassland on this one heavy clay soil display these contrasting features, suggesting the importance of the tillage factor.

Structure studies on the same soil at different dates illustrate the considerable opening effect of contrasting and extreme climatic conditions. A vertical section in May, after a period of drought, shewed a markedly more open structure than one taken in December in the wet winter condition. A demonstration of the effects of deep tillage on the subsoil structure was obtained by contrasting vertical sections of subsoil 8–16 in. deep, one from a plot which had been gyrotilled to a depth of 18 in. and another from an adjacent plot which had received only the normal surface tillages. Fifteen months after the deep tillages, the persistence of the major cavities was still evident, and in addition it was obvious that the clods produced by the operation had weathered to a significant extent, resulting in the appearance of numerous finer channels within them.

So far as these observations have gone, they lead to the conclusion that the permeability of this type of heavy clay is a strong profile characteristic and is a result, in natural conditions, of climatic weathering. It is essentially a dynamic characteristic of the soil and subject to considerable variation. In the natural condition and in grassland the chief factors at any moment are the development of the root systems of the growing herbage and the alternations of wet and dry seasons. Long periods of drought carry the desiccation of the soil and the resultant shrinkage and fissuring to considerable depths. Roots take full advantage of this to extend their range. Irregularities in the rate of penetration of moisture during the onset of the next wet season result in a type of re-expansion of the soil mass which prevents the fissuring from being totally obliterated, so that the permeability of the soil is enhanced for a while. A succession of wet winters or the absence of any pronounced drying-out in the summer tends to obliterate these effects once more. Tillages, with their greater, if temporary, effect on the structure of the soil, have even more influence on permeability. The deeper the tillages, the deeper the permeable layer of the soil. The value of deep tillages in conjunction with the drainage of heavy soils has been on record in this country for many years, and instances of a surprising revival in the efficiency of old and deep tile drains, following on the operation of mole draining, still occur to lend support to these views on the mutability of the drainage properties of heavy clay soils.

It remains to draw attention to the general tendency of ground water to move towards the lower parts of any area, independently of the nature of the underlying material. Where percolation vertically is unhindered, as in deep sand or chalk deposits, the probability is that the water concerned will only reappear at the surface after a long time and at some considerable distance away from and below the place where the rain fell, in the form of perennial springs resulting from the geological structure of the area. If, however, percolation directly downwards meets any obstacle, be it only a layer of slightly less permeability than those above, as frequently happens within the soil profile, it is likely that rain falling at more than a certain rate will produce a surplus which will find it easier to move through the superficial layers of the soil in the direction of maximum fall, but more or less parallel to the surface of the ground. This tendency is most marked in clay and heavy soils generally, and low places are notoriously badly circumstanced as regards drainage conditions.

CHAPTER V

DRAINAGE CONDITIONS IN THE FIELD

From an agricultural point of view the drainage conditions of any site or field depend chiefly on two factors, the permeability of the soil and the physiography of the site. In this connection the permeability of the soil implies also that of the immediate subsoil, and of the geological deposit from which the soil is derived. Over considerable areas of this country, particularly the south-eastern half, there is a close relation between the surface soils and the underlying geological material, but in certain districts complications may arise along outcrop boundaries, or where there are lithological variations within a deposit, or where thin drifts occur. So far as the properties of subsoils or geological materials are concerned, relative to soil-water movement, they may be classified as follows:

Highly permeable	Moderately permeable	Impervious
Sands	Heavy silts	Clays
Gravels	Marls	Shales
Peat	Hard limestone	Igneous rocks
Sandstone	Fissured rocks	Metamorphic rocks
Chalk		
Soft limestones		
Soft ironstones		

This is no more than a broad classification; more detailed information is available for any particular locality in the published geological maps and their accompanying memoirs (see p. 47). These are of immense value in studying particular drainage problems.

In addition to the properties of the soil, the characteristics of the site must be taken into consideration. Its situation relative to the nearest watershed and streams, its liability to seasonal flooding or tidal inundation, the existence of flood plains and alluvial flats, the frequency of ditches, streams, locks and mills may all exert an overriding influence on the drainage properties of the soil as such.

The characteristics of the site decide the ease with which drainage can get away from a field. The reason lies in their influence on conditions in the streams and rivers, the main arteries of any drainage system. Land which lies high or on definite slopes obviously has the advantage of easier get-away for its surplus water. The amount of fall in such circumstances keeps the rivers and streams clear of silt and other debris, the channels are scoured deep, the rate of flow is high and little trouble is experienced. But low-lying or basin-like sites do not possess these advantages, and arterial drainage, or the proper care of rivers, tributaries and major ditches, becomes of supreme importance and confers direct benefit on the whole of the site.

The level of the water in field ditches depends on that in the main drains, and the latter in turn on that in the main streams. If all are in good condition, clear and uninterrupted throughout, the land will be in a satisfactory condition, or can be rendered so by field drains except in times of flood. The incidence of these is intermittent, however, and endures for a short time only, so that the land behind suffers no more than occasionally and for short periods. But in some cases, as in the Fen country, there is so little fall, or the level of the land is such, that the water level in the main drains and ditches can only be kept down by pumping it into the rivers, which run at a higher level. But so long as this is done, the land can be kept satisfactorily drained, and in addition it does not suffer from flooding in the main stream unless either pumping has to be suspended or the bank of the river breaks.

Taking both points into consideration, the drainage conditions inside an area such as a county or similar region may be usefully classified and summarised as follows:

	Elevated, sloping	Slightly elevated, gently sloping	Depressed, basin-like	Flat, low-lying
Impervious				
Of limited permeability				
Highly permeable				
With pronounced spring lines				

The significance of this classification will become apparent after a consideration of the most frequent types of field-drainage problems which are encountered. It will be found that in the areas falling within each subdivision, the same problem recurs with great frequency, and because the fundamental causes are the same the solution or method of dealing with it will be the same in its general outline.

During the early half of the last century, when field draining was being widely practised in this country, much was written on the subject and it is obvious that both principles and practice were the subject of heated and even acrimonious discussion. Much of this arose out of a conscious or subconscious striving after a standard solution of all drainage problems.

Mr Henry Hutchinson, in his *Treatise on the Practical Drainage of Land*, 1844, wrote as follows:

Thirty-one years after the commencement of Deep Draining by Elkington, and when his fame had become extended, we find the Board of Agriculture in this country entered into it very fully, and on the 10th June, 1795, a motion was made by the President in the House of Commons, that a sum of money should be granted to him as an inducement to discover his mode of Draining.

In the year 1796, Mr Elkington's fame had reached the Highland Society in Scotland, when that intelligent body of noblemen and gentlemen despatched a Mr Johnstone, a Surveyor, to watch the progress of Elkington's works, and to report the result to them. After staying a considerable time, and examining the works as far as he was able, he returned to Scotland, and subsequently published his work, describing it as 'Elkington's Plan of Draining'; it, like many other publications, contained many inaccuracies, which the experience of later years has tended to confirm.

Although Mr Johnstone appears in the character of Biographer to Mr Elkington, it does not seem that, like the Prophet of old, he had the 'Mantle cast upon him', nor yet that, during his continuance in the country, he had any portion of the ideas of Elkington himself conveyed to him for his instruction, the general outline of his work being chiefly from his own observations. Nor has it been the case with any Drainers to shew fully, or declare their ideas or reasons to the world. The death of Elkington, shortly after this, put a stop to the further progress of Deep Draining, and left what little knowledge was to be obtained of this important work, in this country, to be obtained from the pen of another countryman.

Turning to the record of the said John Johnstone, thus mercilessly written off, in Chapter II of *An Account of the Mode of Draining*

Land, according to the System Practised by Mr Joseph Elkington, we find:

It is remarkable that the principles on which the draining of land depends, being so great a desideratum in agriculture, should have been so little known or attended to; or that the practice of it, according to these obvious principles, should have been so much confined, while improvements in the other branches of husbandry have been carried almost to the highest possible perfection.

However intricate or abstruse it may hitherto have been considered, even by those who were otherwise well informed in the theory of agriculture, of which it forms the most important branch; yet it will appear, from the following observations, to be founded on circumstances the most plain and rational, and which, when reduced to practice, produce those effects which a simple knowledge of the cause naturally points out.

Wetness in land proceeds from two causes, as different in themselves as the effects which they produce.

It proceeds either from rain water stagnant on the surface, or from the water of springs issuing over, or confined under it. On clay soils, that have no natural descent, wetness is commonly produced by the first of these causes; but in a variety of situations, it may proceed from the latter.

Wetness of land is sometimes occasioned by the stagnation of water in the surrounding ditches, or in some adjoining hollow, where, for want of declivity in the former, and owing to the higher situation of the latter, it oozes out upon the lower ground, and finds its way into the open parts of the soil. This is frequently the case where water is conveyed in a lead, or artificial channel, the land lying lower and adjoining to it, being very often wet from that cause. The remedy for these kinds of wetness is simple, and points out itself.

· · · · · · · · · · ·

Springs, therefore, originate from rain water falling upon such porous and absorbent surfaces, and subsiding downwards through such, till, in its passage, it meets a body of clay or other impenetrable substance, which obstructs its further descent, and here, forming a reservoir or considerable collection of water, it is forced either to filtrate along such body, or rise to some part of the surface, where it oozes out in all those different appearances that are so frequently met with. This is evident from the immediate disappearance of the rain water, as it falls, on some parts of the ground, while it remains stagnant on others, till carried off by evaporation; and from the strength of springs being greater in wet than in dry seasons. Hence, after incessant rains, they are observed to break out in higher situations, and, as the weather becomes drier, give over running out, unless at their lowest outlets. The strength of springs also, or quantity of water which they issue, depends chiefly on the extent of high ground that receives and retains the rain, forming large reservoirs, which affords them a more regular supply. Thus, bog-springs, or those that rise in valleys and low situations, are much stronger, and have a more regular discharge, than those which break out on higher ground, or on the sides of hills.

. . . .

As the whole depends upon the situation of the ground to be drained, and the nature and inclination of the strata of which the adjacent country is composed; as much knowledge as possible must be obtained of these, before the proper course of a drain can be ascertained, or any specific rules given for its direction or execution.

Mr John Johnstone would appear to have been the possessor of a sound knowledge of the fundamentals of the science of ground-water movements, even if he did not give satisfaction in his effort to put into simple phraseology the secret of Joseph Elkington's art. But perhaps he was damned by his connection with the Board of Agriculture, even in 1801.

Impervious Land in Elevated Situations and on Slopes. One is frequently faced in farming with the anomaly of standing water on high land. It is particularly common on the Boulder Clay areas of East Anglia and the East Midlands. Heavy impermeable soils, formerly almost invariably worked in narrow high-backed lands, lie wet for a goodly part of the winter, especially between December and March, with a fair proportion of the ground actually submerged. Where the area is flat the problem exists in its most acute form. The trouble is due first of all to the clay subsoil, through which water cannot pass, and to the lack of fall wherewith to get it away from the land to the ditches, and from the ditches to the streams. In such circumstances the siting and grading of the land drains and ditches are the vital points in any scheme of drainage. Care and maintenance of these works are most essential in order to keep the land as dry as possible. The cost is heavy, for in such circumstances the ditches in particular are numerous and deep (see Fig. 16).

Impervious soils on slopes give less trouble in draining. The same problem of surface water-logging and widespread pools is common, however, but in these cases the solution is more easy; ditches and drains alike are less difficult to site and maintain. Where the former run down the slopes they are frequently self-scouring and only need attention on rare occasions. The latter are easily sited and can be placed so as to give the maximum benefit. It is, indeed, easy in such circumstances, to give a drain an excessive fall, to its undoing. While ditches down the slopes may give little trouble, it may be stressed that those across the slopes, while often extremely important, are more liable to go wrong. Their importance lies in their strategic position for intercepting any water moving down the slope. In dealing with the water from above, they are essential to the land below. Running across the fall, they derive less advantage from it, are more liable to silt up, need more constant attention and are a greater potential source of trouble.

Further considerable areas of this country, more particularly in the north and west, come within the category of impervious materials on high-lying sites. They are impervious owing to the nature of the rocks of which these hilly and mountainous regions consist. They are also endowed with rainfalls which are much higher than the average, so that surface water-logging persists over a much larger proportion of the year. This has its inevitable effect on the vegetation and soil conditions which are there found. A layer of peat accumulates, in which surplus water lies as in a sponge. The commonest way of dealing with this class of land is by a system of shallow ditches. They are in effect surface cuts and are termed sheep drains.

Impervious Land on Gentle Slopes or in Depressions. Where impermeable soils lie on gentle slopes, the drainage problems and their solution are similar to those already outlined, the only difference being the matter of fall. But it is in depressions, or basin-like areas, or in flat, low-lying country that impervious soils are found in the worst case. Here, everything is against them, and to the burden of the cost of field drains is added that of arterial drainage. To begin with, because of the lack of fall, it is not easy to get the water off the land into the ditches. Even mole draining, that great stand-by of heavy-land farming, is ineffective unless backed by a liberal and careful provision of tiled mains; and it is useless doing this if the minor streams and the main rivers are not in tip-top order to provide an uninterrupted exit for the run-off.

Permeable Soils in Low-lying Sites. If the soil and subsoil in these circumstances are freely permeable, e.g. as in the case of peat or gravel, the ditch, which generally surrounds the field entirely, may be the only drain required. Provided it is kept in order and its contents are steadily drawn away, i.e. if the arterial system as a whole is efficient, the field will be kept dry. If, however, the subsoil is only moderately permeable, as in the case of river alluvium, silts, or marls, field drains will be required in addition to keep the ground-water level under control. The less permeable the soil and subsoil are, the more intensive the field-drain system which is necessary. It usually takes the form of occasional tile drains running from one ditch to that opposite, on the level, at or a little above the bottom of the ditch.

Permeable Soils in Elevated Situations or on Slopes. Whereas percolation is almost wholly confined to the surface soil in the case of clay formations and as a result is followed very quickly by movement down the slopes in the surface soil, the same is not the case in open soils, as has been described. Percolation proceeds vertically until arrested by an impermeable stratum, on which it then accumulates as a rising water table. If this stratum, of clay for instance, lies near the surface and outcrops along a slope, the presence of the water table will soon become apparent as a spring line with waterlogged ground both above and below it. The distance to which this condition extends above the spring line depends on the nature and inclination of the underground clay surface, the slope of the surface of the ground, the extent of the gathering ground behind the spring line, the rainfall and the time of year. The water-logged area will spread both up and down the hill during the winter, according to circumstances, and will tend to disappear in the summer. Some conception of the scale and variations of this simple case is given by the following examples. Fig. 11 represents the cross-section of a gravel deposit, lying on a stiff impervious clay in an elevated position, with free-working light soil between *A* and *B*, flanked by clay land on either side. The dotted line shews the position of the water table at its highest, generally in February. In this actual case, the gravel area is 400–500 yd. across and 20 ft. at its greatest depth. Ponds occur at *A*, a spring line and an interceptor ditch are found at *B*, most of the land between *A* and *B* is tile drained, and occasionally water stands in the depression at *C*. Although the catchment of the area of the gravel is small, the deposit is so thin that the water table produced in it comes up towards the surface rapidly and often enough to be a nuisance and to necessitate field draining over a goodly proportion of its total area.

Fig. 12 shews the nature of the deep water table in the Chalk formation relative to its escarpment as it occurs in Cambridgeshire. *A* is a point in the bottom of the river valley, *C* another on the Chalk ridge on the Essex-Cambridgeshire boundary. The water table assumes the form and position to be expected in a substantial mass of readily permeable material with a sloping surface, i.e. the surface of the water table reflects the slope and irregularities of the land surface to a slight extent, with the depth of the water table from the surface rapidly increasing with the height of the latter above o.d. It will be noted that in the valley bottom where

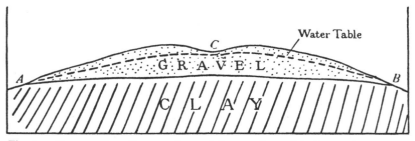

Fig. 11. Cross-section of a gravel area lying on clay, with the resulting water table. *A* and *B* are the positions of ponds, springs, or interceptor ditches. *C* is a depression which is often wet and occasionally the site of a temporary pond.

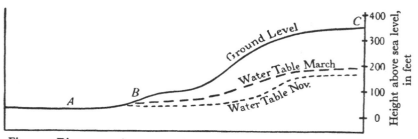

Fig. 12. Diagrammatic cross-section illustrating the deep water table in the Chalk ridge of East Anglia. *A*, the valley in the Gault Clay. *B*, the lowest Chalk springs. *C*, the crest of the Chalk ridge.

the land surface is only about 60 ft. above sea level, the water table is very near the surface and field drains are often necessary. From *B* to *C*, the water table, although rising in height above sea level, gets farther and farther below the ground until on the hills at *C*, 400 ft. above sea level, the water table is some 200 ft. below ground, but still 200 ft. above sea level. It is worth while noting also that the water table is at its lowest in November and highest in March. Many people are familiar with certain localities in Chalk country, where springs break out occasionally in March and convert otherwise dry and cropped land into areas of running water, at intervals of several years.

Yet another variant is illustrated in Fig. 13. This is a most mystifying phenomenon when it is encountered in real life. A wet belt of land occurs along the side of a slope, the water can be seen issuing from the hillside as a spring line, but lower down it just as suddenly disappears and below that the land is dry. If there happens to be a ditch running down the slope within the area affected, it also is found to be wet or running within the same belt,

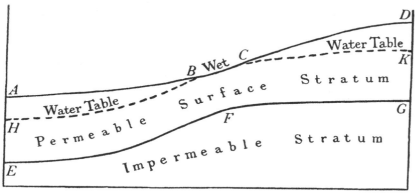

Fig. 13. Diagrammatic cross-section of a hillside with a belt of wet land between certain levels. *ABCD*, the ground surface. *EFG*, underlying impermeable stratum. *HBCK*, the water table. *BC*, the wet belt.

but above and below the bottom of the ditch is dry. Various conditions underground can produce this, but they all have one point in common: a layer of lower permeability comes nearest to the surface of the ground at the wet belt with the result that the water table there is forced up to the surface. It may be a chemical hardpan, a vein of clayey material, or a bump in an underlying clay formation. In Fig. 13, *EFG* is the surface of such a clay deposit, *ABCD* the ground surface, *HB* and *CK* the water table and *BC* the wet belt, where, due to the bump at *F*, the water table reaches the surface and appears above it, issuing at *C* and running down to *B* and then disappearing below the surface once more.

Major Spring Lines. The more important spring lines in the country become aggressively apparent in February or March in most years and particularly after wet winters. The fields thus affected can be marked off on the 6 in. O.S. map after studying the corresponding Geological map and subsequently confirmed by inspection on the ground. There is no mystery about these spring lines. Wherever the boundary between the outcrop of a permeable formation and an underlying impermeable one occurs there will be found a belt of land liable to wetness unless it has been adequately drained. The simplest case, which has already been discussed, is that of the edges of a bed of sand or gravel over clay. The same result accrues if the difference in permeability between two adjacent strata is only small, provided the less permeable layer is underneath. A striking example is associated with the Chalk formation. This substantial deposit, hundreds, and in some places

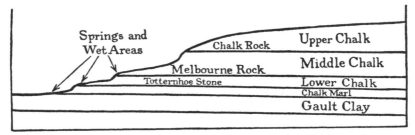

Fig. 14. Cross-section shewing the divisions of the Chalk formation, the basal rock beds of each, and the resultant spring lines.

thousands, of feet thick, consists of three main divisions, called respectively the Upper, Middle and Lower Chalk. The basal bed of each is harder, denser, more compact, and so less permeable than the bulk of the deposit above it. That of the Middle Chalk is known as the Melbourne Rock, that of the Lower Chalk the Totternhoe Stone. These beds are only a few feet thick and their outcrops on the map shew only as thin lines, but it will be found that the fields through which these two lines pass are quite frequently characterised by the presence of springs and wet areas (see Fig. 14).

Similar circumstances are encountered along the edge of gravel deposits lying on clay formations; along the boundary of the Chalk Marl or Upper Greensand with the Gault Clay below; along the boundary of the Lower Greensand with the Kimeridge Clay or the Oxford Clay; along the boundary of the Upper Calcareous Grit with the Coral Rag below; and so on.

CHAPTER VI

THE USE OF PUBLISHED MAPS IN FIELD-DRAINAGE WORK

In the last decade or two before the war, since field-drainage work has ceased to be part of the routine on many of our farms, and when skilled drainers have become few and far between, great deference has been paid to hearsay evidence, much of it second if not third hand, of past drainage conditions and work carried out in particular fields and localities; but the factual records of the Ordnance and Geological Surveys have received scant attention. Even now, few realise what a wealth of information exists in published maps with respect to the drainage conditions of any field, farm or estate. Nowadays, however, the requirement by the Ministry of Agriculture that a 6 in. plan of proposals for ditching work shall accompany applications for grant, and that a 25 in. plan is to be made of tile-drainage work carried out under a grant-aided scheme, is compelling much more attention to the published maps of the Ordnance Survey.

One or more complete sets of the 6 in. and 25 in. O.S. maps is part of the equipment of every War Agricultural Executive office, and its staffs have frequent occasion to refer to them, e.g. to the former to locate a farm, to the latter to get the number and area of any individual field.

In drainage work the maps can be of even greater use, especially if studied in conjunction with those of the Geological Survey, whose work is based on, and accurately recorded on, the 6 in. O.S. map, and subsequently published in the 1 in. form.

The inception of the Government schemes for assisting mole draining in January 1940, and the extension of similar aid to ditching and tile draining in July 1940, has thrown on to War Agricultural Executive Officers, their Drainage Officers and District Officers, a considerable amount of work in examining proposed schemes, advising on drainage problems, and devising schemes under all three headings. Fairly large acreages must be considered and dealt with in a short time, and to accomplish this without loss of efficiency is obviously desirable. It is not for a moment suggested that drainage problems can be solved in an office, but much valuable time can be saved by studying the available recorded data before inspecting the site. Armed with this data, the inspector can the more effectively and quickly come to a decision on the spot. It frequently happens that vital factors, accurately recorded in the existing maps, are not at all obvious in the field, and can only be elucidated by prolonged and laborious investigation.

The drainage of any field depends on two main groups of factors, those of the site and those of the soil. Those of the site are its position relative to surrounding land, whether high lying, on a slope, or in a valley bottom, its nearness to and access to a main stream, and its liability to flooding. Those of the soil are decided by the nature of the soil, its subsoil and the deeper strata, particularly their permeability to water.

The following is a summary of the useful data which can easily be gathered from published maps:

Six-inch Maps, O.S. All objects shewn are to scale, so that with the eye 1 ft. away from the map the appearance of the ground or object is the same as it would be actually from a height of 2 miles.

Frequent levels of the surface above sea level are shewn, mostly along roads or the banks of streams. It is, however, necessary to bear in mind the difference between a Bench Mark (B.M. on the map) and a surface level. The former refers to an actual mark on a building, wall or post.

Contour lines are accurately drawn for heights of 50 ft., 100 ft. and at intervals of 100 ft. up to the 1000 ft. level. Above this they are given for intervals of 250 ft.

By means of the surrounding surface levels and contours the general direction of fall of any field can be deduced from the map. Its irregularities, of course, can only be determined on the ground.

The map gives the most accurate idea possible of the shape of any field, a bird's eye view in fact, and with the help of the scale or an ordinary ruler, an accurate assessment of any of its dimensions is possible. These factors, of course, are of vital importance in deciding on the lay-out of a drainage system with respect to the length of the minors and the length of the mains.

Streams and major ditches are shewn by a double line and the direction of flow by means of an arrow. Farmer's ditches, however, are not specially indicated and there is no means of distinguishing them on the map from other forms of field boundaries.

Other water features of diagnostic value are ponds, lakes, moats and springs. Frequent ponds, one or more in each field, are a sign of clay country or places where a clay stratum lies near the surface. A line of ponds often betokens the boundary between the outcrop of a clay formation and a higher-lying more permeable material such as gravel or sand. The simultaneous and frequent occurrence of all these water features means a natural high-water table throughout the neighbourhood.

Areas liable to flood are often indicated on the maps, to the enlightenment of those not actually acquainted with any locality, as also are swamps. Deciduous woods and trees in some districts, by their position and frequency, indicate where water comes nearer to the surface.

Gravel, clay, chalk and marl pits are all marked as such and frequently give some indication of subsoil conditions in their immediate neighbourhood. On the site they are worth inspection for indications of ground-water level.

Twenty-five inch Maps, O.S. It is important to remember that these maps are not accurately named. The scale is 1/2500 or 25·344 in. to 1 mile, so that the ordinary ruler cannot be employed for the accurate measurement of distances on these maps without a correcting factor.

They contain all the information given in the 6 in. maps. Many more 'spot levels' are included, however. In addition, the O.S. number of each field and enclosure in any parish is given, together with its exact area. These items are most important in county drainage work and the chief drawback of the 6 in. maps is that they are not given thereon. It is, however, easy to transfer them from one to the other for use in the field.

One-inch Drift Maps, Geological Survey. The information to be obtained from these concerns the drainage properties of the soil itself. In the south and east of England there is a close connection between the geology and the nature of the surface soil, particularly its drainage conditions. From the Drift map and its accompanying memoir can be gleaned a detailed description of the subsoil material, whether alluvial (peat or silt), gravel, sand, clay, chalk, limestone, sandstone, or hard rock. Thus it is easy to see whether one is dealing with a permeable or impermeable material, whether the site is likely to be influenced by underlying material, whether a spring line is concerned, and if so, its real position. The boundaries shewn in 1 in. Drift maps are in the main very accurate. Furthermore, since in respect of many of the physical characters shewn, these maps are exact replicas of the 6 in. O.S. maps, it is possible without much effort or equipment to transfer the boundaries from the 1 in. Drift map to the 6 in. O.S. and so get accurate information about any individual field. A change in the surface geology is often of first importance in drainage problems, and the method of dealing with them should take due account of it.

Sufficient has been said, it is hoped, to indicate the way in which maps can help in field-drainage work. Their utility increases as the local experience of the individual grows. The same circumstances in any field lead to the same drainage condition or problem and the method of treatment should be the same, so that the labour of dealing with large numbers of individual fields within any district is correspondingly lightened, if the maps are used for a preliminary study, to be followed by verification on the actual site.

(The contents of this chapter constituted a leaflet circulated in November 1940.)

CHAPTER VII

THE INVESTIGATION OF FIELD-DRAINAGE PROBLEMS

From what has been said on the use of maps, it can be seen that a fairly clear and concise mental picture of any given field can be formed before visiting it. With this in mind, or better still on paper, the salient features can be checked on the ground the more quickly and the investigator's judgment confirmed with the greater celerity. In walking over the field, the observer should give special attention to the ditches and their condition. Neglect of these in past years may be the sole cause of any trouble which exists. The extent to which a ditch has silted up can be gauged easily in winter by means of a walking-stick, though cases do occur where that useful article is not long enough to probe through to the original floor of the ditch. In summer a spade may be necessary and one's judgment based on the appearance of the material turned up. It is an easy matter to distinguish between silt deposits, falls from the sides and undisturbed earth.

The depth and condition of culverts under gateways and roads or driftways are important. Very often, culverts lie deeper than the ditch bottom, due to neglect of the latter; they are very rarely too deep, however. Sometimes they are obviously too shallow, but in such cases it will usually be found that there is a deeper one which has either been lost or destroyed.

Signs of the existence of submerged outfalls should be looked for. There is little land in this country, if it has ever been in cultivation and is now in need of draining, which has not been drained at some time in the past, when agriculture was more prosperous; and lots of old drains are daily being uncovered and found to be capable of running, when freed. In winter, some of these outfalls reveal their position by a boiling up of the water in the ditch; when the water is turbid, the turbulence alone is sufficiently marked to catch the eye; when it is clear, a dancing fountain of small gravel or coarse sand particles in the water of the ditch indicates the whereabouts of the outfall. In summer, if the ditch is dry, the sorting of the gravel and sand above the outfall is frequently a reliable indicator.

It is not only the ditch which receives the water of a field which is important. If a field lies on a slope, its side and top ditches act as interceptors, protecting the field against water from higher adjacent land. They are especially important in open soils, where they function as drains in themselves. The continued neglect and filling-up of such ditches is often followed by the spread of a wet area down the slope below. In heavy land their main function is as carriers, but if neglected or obstructed they may easily overflow and drown the field below during 'fill-dyke' time. The utilisation of ditches as drinking places for stock is a potent cause of trouble. The deliberate damming of a ditch for this purpose is a most reprehensible practice.

The recognisable indications of wet land vary with the season of the year and are to be looked for, sometimes in the soil, sometimes in the herbage on it. The bogging of tractors, horses, or men in soft wet ground is obvious enough, as also is the occurrence of standing water. In light land, such evidence is most likely to be forthcoming in February or March; in heavy land the trouble is frequently obvious enough in November or December. So soon as the drying winds of March exert their influence, areas of high water table begin to shew up in the dark colour of the moister surface soil, the result of capillary rise of moisture from below and the non-drying-out of the surface.

Certain weeds favour moist land and wet places. Such are rushes, sedges and horsetails; silverweed, coltsfoot and wild onion (though these also characterise heavy land); mosses, orchids and yellow rattle, particularly in grassland; cowslip, butterbur, lady's smock, persicary and ragged robin; meadow sweet, willow herb and hemp agrimony (although these last are common at ditch sides, they do invade wet places if these are neglected).

Wet land is notoriously late in every respect, and this is reflected in the development of the herbage on it. Growth begins later and is more feeble. In a field with wet patches in it the contrast in the herbage on them and on the surrounding dry areas easily catches the eye of the observer. Among the crops of arable land, winter-sown cereals reflect excessive wetness by marked yellowing of the leaf and dying off of the plant. Brussels sprouts, especially towards the end of their tenure of the ground, shew up the wet patches by their colouring, browning of the tops and early death. On the other hand crop variations will often reveal the exact line of existing drains. A thin band of darker colour and enhanced growth in the crops such as mangolds, sugar beet and cereals, particularly during the summer in periods of drought and towards harvest time, often demarks the line of a tile drain.

While thus exploring the position by eye, a spade should be brought into action, aided occasionally by its companion, the soil auger, which should be of the screw type and of a fair length. By this means it is possible to examine the soil in vertical section for the actual presence of water or its signs. The vagaries of the water table have been indicated; for the greater part of the year, on many sites, there may be no water near enough to the surface to be observed, but between December and April it is liable to appear with disconcerting suddenness. This fact has been forcibly brought home to thousands who made themselves dug-outs in the autumns of 1938 and 1939, to many who sited and constructed earthworks in the summer of 1940, and in fact to all those who of late years have taken to a troglodyte existence. Yet the signs of wetness and fluctuating water tables are always present in the soil and easily found by those who take the trouble to look for them. The drainage conditions in any soil are one of the chief factors responsible for its appearance in profile; and this is where the spade makes its contribution. Holes should be dug deep enough to expose the subsoil and the profile changes in their entirety. As a rule, to do this necessitates a hole 2 or 3 ft. deep. Soils which are naturally well drained have full colours throughout the profile, uniform in each horizon and changing with depth only in respect of brightness. Such soils may be reddish, brown or yellow. Badly drained soils are often associated with greys, blues or greens, especially in the subsoil. The colour is not uniform in any horizon, but characterised by mottling of a yellow, orange or brown colour in the grey, blue or green background. This mottled appearance is due to alternations of oxidising and reducing conditions resulting from the fall and rise of the water table. These changes affect the iron compounds in the soil and produce the streaks or patches of colour on the outside of the structural units of the soil, or in amongst the sand particles in the case of the more open soils. One other sign of a fluctuating water table is to be found in the form of grains or concretions of manganese and iron compounds varying in size from that of mustard seed to that of peas, scattered along a certain horizon of the soil. They are purplish or black in colour and fairly soft. They seem to occur in the higher parts of the zone of soil subject to the fluctuations of the water table. During those parts of the year when the water table is near the surface, the spade is useful to ascertain to what extent the problem is one of top water, bottom water, or both. With the auger and spade, also, it is possible to probe cases where an underlying impermeable stratum is the cause of trouble, especially if it outcrops on the surface of sloping ground. But there are cases where this kind of trouble is due to strata so deep that they are beyond the range of ordinary spade work.

A burst or broken drain often gives itself away by an outbreak of water in one place, an overground flow for a distance, and the disappearance of the water underground once more. The marks of such events, too, persist on the surface for a long time. In very wet spells, in spring, the surface flow or drift of water over stubbles or grassland is a useful guide to the whereabouts of important former drain lines where these have gone out of action for some reason or other.

The study of drainage problems is fascinating for its own sake; so much that is involved in their solution is hidden from view and must be ferreted out by methods of the kind indicated above. Time spent in reconnaissance is seldom wasted; this principle applies with especial force in drainage work. Accordingly, the approach to a drainage job should be as thorough a survey as is possible, making use of all known, recorded and discoverable facts. If the drainage conditions can be correctly deduced, the cause of the trouble will be apparent and measures can be soundly planned to deal with it. To deal with all field-drainage problems on empirical lines or by rule of thumb methods is unsatisfactory; any failures which result can only bring into disrepute practices which are sound in themselves and effective where applied in the appropriate circumstances.

CHAPTER VIII

DITCHES

Of the various items of drainage under the direct control of the farmer, the ditch is the most important; and there is no doubt that of all the phases of drainage work being undertaken at the present time, that of ditch cleaning is producing the greatest all-round benefit. The term ditch here is used to describe mainly those waterways bordering or surrounding individual fields, the concern of the farmer himself. There are of course some which are of major importance, and deal with the water of bigger areas of land and which are the concern of public drainage or local government authorities.

In the main, ditches are of two kinds, natural and artificial. It is obvious that even before land was enclosed, there was a natural drainage system in the shape of existing rivers, streams, and all their minor branches, mostly perennial, but some running seasonally only. They can be recognised at the present time by their irregular courses and the greater number of trees along them: they constitute the backbone of the ditch system and many of them are so-called public drains (see Fig. 15). Some, however, are, from the drainage point of view, ordinary field ditches. In contrast to these there are many more, the work of men's hands, sited and constructed with the object of dealing with the drainage of definite parcels of ground. Such are obviously characterised by running fairly straight from point to point and joining each other more or less at right angles. It is frequently found, especially when they lie across the slope of the ground, that their position coincides with a change in soil type. This often connotes a change in drainage conditions, so helping one to a realisation of the importance to a field, in many cases, of the ditch along the top side. A ditch may fulfil one or more of several functions; it may be a drain itself and as such draw water from the land above it and intercept underground flow, thus protecting land on its low side; or its purpose may be to carry water from elsewhere through the land on either side. Ditches down the slope mainly function as receivers or carriers; those across the slopes do this and also act as interceptors to a certain extent.

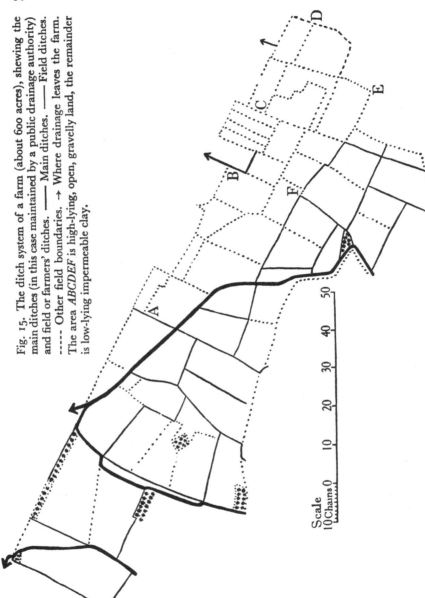

Fig. 15. The ditch system of a farm (about 600 acres), shewing the main ditches (in this case maintained by a public drainage authority) and field or farmers' ditches. ——— Main ditches. ——— Field ditches. ----- Other field boundaries. → Where drainage leaves the farm. The area *ABCDEF* is high-lying, open, gravelly land, the remainder is low-lying impermeable clay.

Scale
10 Chains 0   10   20   30   40   50

The incidence and importance of ditches varies with the class of land. High-lying and free-draining land, such as, for example, much of the chalk and limestone country and many sandy areas, is characterised by a marked absence of them. But in flat, or low-lying land, or areas surrounded by hills, they are indispensable. The anomalous conditions on high-lying clay land have already been commented on; in some such districts the expense of ditching is a serious matter. In certain districts, of flat land with permeable soils, the ditches may be the only drains necessary, provided they are kept in order. The black Fen lands of the eastern counties are a case in point; here the fields are wholly surrounded by ditches. The importance of ditches in the farming economy is illustrated in Table IV.

TABLE IV. *Shewing amount of ditches employed*

Locality	Drainage conditions	Area acres	Length of ditches chains	Length of ditch per acre chains
Coveney (I. of Ely)	Fen peat, 2 ft. o.d. (flat)	68	287	4·2
Purls Bridge (I. of Ely)	The Washes, 2 ft. o.d. (flat) floodland	260	633	2·4
Burnt Fen (I. of Ely)	Fen peat, at o.d. (flat)	321	661	2·1
Lake's End (Norfolk)	Fen silt (light), 9 ft. o.d. (flat)	265	442	1·6
Seething (Norfolk)	Boulder clay, poor falls, 200 ft. o.d.	110	350	3·2
Girton (Cambs.)	Gault clay, 50 ft. o.d.	325	464	1·4
Boxworth (Cambs.)	Boulder clay, good falls, 200 ft. o.d.	370	400	1·1
Girton (Cambs.)	Gravel and loam, 80 ft. o.d.	130	64	0·5

Not only is there a tremendous range in the length of ditch in a given unit of area, but the types and functions of these ditches shew marked contrasts. Thus, in the Fen peat country, most of the field ditches are roughly 3 ft. wide by 3 ft. deep and during the winter fairly full of water. In this type of land, the surface is at or below sea level, the water table is high, and the aim is to get it down if possible to 3 ft. or so below the surface. To accomplish this

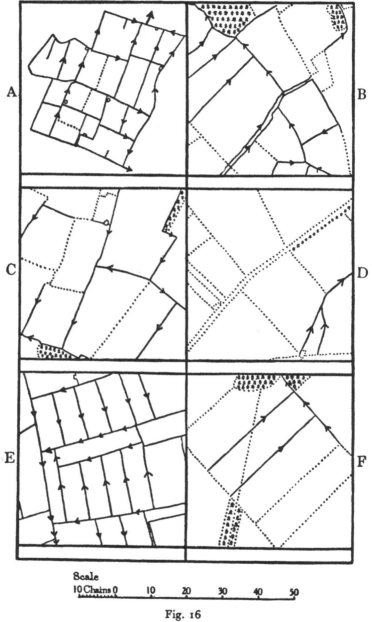

Scale
10 Chains 0   10   20   30   40   50

Fig. 16

the water has to be pumped out of the ditches, so these in winter are generally carrying a fair load. In the Fen silt areas the ditches are much deeper, many of them 10 ft. wide at the top and tapering to 2 ft. at the bottom, which is 7–8 ft. deep. Their sides are well covered with vegetation and one rarely sees more than a foot of water in the bottom. In these circumstances it will be found that the subsoil is mostly fine sand, uniform to a fair depth, and the aim is to get the water table well down during the summer and as far as possible prevent it from rising near to the surface during the winter. This the deep ditches, with occasional deep drains across the fields, succeed in accomplishing. In the case of ordinary farm land, the burden of ditches is lighter; the better the drainage conditions, the fewer seem to be necessary. Their size, depth and shape depend on the conditions of the site. Some fields, too, are burdened with a greater length of ditch than others and there is a tendency to view the whole subject in terms of the number of chains of ditch per acre of the smallest field on the farm. On the average, farm ditches run at a chain or a chain and a half per acre (see Fig. 16).

In re-conditioning work, a careful reconnaissance of the whole unit concerned is well worth while, in order to decide on the relative importance of the individual ditches with a view of the order in which they should be dug out. They should be considered as a miniature river system with tributaries. Obviously there should be one or more functioning as main channels, with increasing

---

Fig. 16. The relation between the drainage characteristics of site and soil and the ditching problem. ——➤—— Ditches; ······· other field boundaries.

A, Flat land, 200 ft. above o.d., heavy Boulder Clay soils in south-east Norfolk. Small fields, 6–12 acres, wholly surrounded by ditches.

B, Mostly undulating land, 200 ft. above o.d., heavy Boulder Clay soils in Cambs. Fields 10–40 acres, ditches as a rule on three sides of the field, some of them being natural self-scouring streams.

C, Partly sloping, partly flat land, 80–190 ft. above o.d., heavy Boulder Clay and Oxford Clay soils in Hunts. Fields 20–45 acres, ditches on two or three sides of the fields.

D, Land with a definite slope to the stream, 130–200 ft. above o.d., open permeable Chalk soils in Cambs. Fields 20–50 acres, mostly without ditches.

E, Flat land, 2 ft. above o.d., Fen peat soils in the Isle of Ely. Fields 6–15 acres, surrounded by ditches.

F, Land in a shallow depression, 60–70 ft. above o.d., open permeable Chalk soils in Cambs. Fields 30–40 acres, ditches on two or three sides of the lowest fields.

loads as they proceed on their way through the farm. With the help of a map (6 in.) it is possible rapidly to arrive at an idea of the area served by the ditch at any point and so of the load of water it may have to carry. This in its turn should influence the size of the ditch.

The matter of continuity in a ditch and freedom from obstruction is important. It can be overstressed, however. How often one hears the statement that because the man below neglects his ditches therefore the man above has little option but to do likewise. There is no gainsaying the soundness of this argument in the case of flat or almost flat land. A case in point was that of some 80 chains of main ditch, serving 150 acres of land, in thirty small fields. The whole length had a total fall of 18 ft. or 1 in 300; there were four culverts blocked or hopelessly collapsed. The thirty fields were in the occupation of a number of occupiers, of whom one or two were eager for improvement; but in the absence of agreement amongst all of them, the position was one of deadlock and nothing was being done. In some cases, on the other hand, this argument is an excuse, rather than a valid reason for inaction. Wherever the general fall exceeds, say, 1 in 200, it no longer holds good. Suppose, for instance, that there is a hold-up of this sort and the ditches of the farm up above are put in order. The drainage will quickly get down to the bottom of the farm at all events, where it may possibly be harmful to a few acres for longer than usual; but the bulk of the farm will undeniably benefit considerably. The water will to a certain extent force its way out through the obstacle, and in doing so may call attention to the need for action farther down.

The maximum rates of run-off of drainage water so far recorded and published for this country are 1,500 gallons per acre per hour from clay land and 500 gallons per acre per hour from open soils. As it happens these records were made in the eastern counties and it is possible that higher rates occur elsewhere, especially on heavy land. The consideration of a few individual cases with the help of simple formulae for flow in open channels shews that a ditch 1 ft. across the bottom, with vertical sides to a height of 1 ft. from the bottom, on a gradient of 1 in 200, would cope with the flow from 54 acres of mole-drained clay land when running at a depth of 1 ft. A stream 3 ft. wide by 2 ft. deep, on a gradient of 8 ft. in a mile, would cope with the flow from 300 acres of such land or

1,500 acres of open soil served by tile drains, so that carrying capacity for ordinary farm requirements would not appear to be difficult of achievement:

$$V = \sqrt{\frac{a}{p} \times \frac{3h}{2}} \text{ and } Q = av,$$

where V = the mean velocity in feet per second,
 a = the area of cross-section of the ditch in square feet,
 p = the wetted perimeter in feet,
 h = the fall in feet per mile,
and Q the rate of flow in cubic feet per second.

In real life, unfortunately, the problem is not so simple as this. As in many other human activities, and the estimates connected with them, allowance must be made for contingencies as well as other variables. For instance, any unevenness of grade or irregularity in shape tends to obstruct the flow of water and cause it to pile up at certain points. To cope with localised overloading, a greater area of cross-section is necessary. Erosion, falls of the sides, and so on, continually diminish the capacity of the waterway, which accordingly must be big enough to prevent resultant overflow. Further, the ditch being the receiver for the mains of field-drain systems which are usually at least 30 in. and often as much as 5 ft. deep at their outfalls, it must have sufficient depth to allow these outfalls to run freely without being perpetually or even frequently submerged.

Ditches with vertical sides could only be made in more or less solid rock; in loose materials the sides must be sloped if they are to possess any stability at all. Both the natural angle of repose and the erodibility of the material are involved. From the point of view of the stability of the soil in which the ditch is dug, its sides should have a slope of 1 in 1 in clay, 1 in 2 in loam, and 1 in 3 in sand. So it is found that steep-sided ditches are more stable in clays than in sands, if they are dependent on the amount of batter only. Generally speaking, it will be found that ditches are of such a cross-section that the width at the top is a foot or two greater than the depth, and the width of the floor 18 in. or more, according to the load the ditch is called upon to carry. It is, as a matter of fact, the herbage of the ditch side which makes the biggest contribution to its security. 'When ditching was ditching', as the old hands say, it was common to turf the sides of the ditch so that the grass roots could penetrate them, take a firm hold and thus consolidate the whole, while also protecting it from the disintegrating effect of rain and frost. This important contribution of the ditch flora should be kept in mind when a ditch is being cleansed or re-conditioned. Unless the enlarging or deepening of the ditch demands it, care should be taken not to slice off more of the sides than can be avoided; to trim them down for appearances' sake is to be deprecated. On the other hand, it is necessary to keep the actual waterway clear of weeds which may grow in the bottom and also of the dead stems of those which grow on the banks. This is the essence and object of the annual brush-out. If this is omitted even once, it is amazing how much the efficiency of the ditch is impaired. For the ditch sides and the berm each year produce an astonishing growth of weeds, fine upstanding specimens of every variety, often entangled amidst trailers of blackberry, dewberry and briars. On the floor of the ditch, if it runs throughout the year, such vigorous individuals as brooklime, water crowfoot, water mint, hemp agrimony and others flourish amazingly. All of them in autumn and winter add their quota of debris to obstruct the ditch. A single year's growth is enough to raise the water level in the bottom of the ditch by as much as 6 in. in a chain. This means, of course, that many outfalls are submerged at once, and in another year or two they become silted up and lost to sight. The hindrance to the flow of water increases the rate of silting; leaves blow in, and soon the ditch contains a soft mass of mixed silt and organic matter to a depth of one or more feet.

One side of a ditch often carries a hedge; this also, by the development of its roots, stabilises that side of the ditch, though the roots can become a nuisance in ditch cleaning. Unfortunately, continuous neglect in many cases has enabled the hedge to grow under the ditch and establish itself firmly on the other side also; such cases constitute some of the toughest jobs of reclamation work. It is frequently the existence of a hedge or fence, together with occasional trees, which hampers the use of machinery for ditching purposes. On the score of cost and convenience, and in overcoming the shortage of labour, excavators have many advantages, particularly in constructing new ditches or reclaiming old ones, where there are no obstacles to their approach, but it is to be feared that most ordinary farm ditch work must continue to be dealt with by hand.

Plate III *a*. 'Where is the bottom of this ditch?' (*See p.* 49.)

Plate III *b*. The blocked culvert. The top of the arch is just visible above the water in the ditch; this approaches from behind the camera and turns sharply to the left under a driftway; the reflection of the sky is seen in the water. (*See p.* 49.)

Plate IV b. Cleaning operations, cutting back the hedge, trimming the sides and 'brushing out' the bottom soon have it functioning once more. (By permission of *The Farmer's*

Plate IV a. The neglected ditch. Hedge and weeds have combined to conceal it from view. Unless cleaned up, it will soon cease to function, as also will the drains running into it. (By

As far as possible a ditch should be straight or run in smooth curves. Junctions obviously are a difficulty, as also are the corners of fields, but here again the importance of the individual channel and the volume of water to be dealt with should influence the lay-out. Anything which can lead to obstruction should be avoided as far as possible.

Fall or gradient is important chiefly because of its influence on erosion and silting. The swifter the flow, the greater the wear on the banks and floor of the ditch; and wherever the current slacks off there is bound to be some deposition of silt. Accordingly, as far as possible, an even grade should be established, in bottoming out the ditch. The small amount of trouble involved in the erection of sighting rails in work of this description will be repaid many times in the efficiency of the resulting ditch. In actual fact there are places in all ditches where silt accumulates at every run of the drains, and these should receive special attention. Such points occur just below drain outfalls, at the junction of ditches, at places where the grade falls off, and where the ditch broadens out.

In bottoming out ditches one will rarely err in restoring them to their original condition; all dark-coloured silt, earth and humus should be removed until the raw subsoil is exposed once more. Culverts under tracks and gateways are a sound guide to the correct base line. Unfortunately cases are not infrequent in which the original culverts have been lost sight of altogether through neglect, and the present channel may be merely an *ad hoc* make-shift of a later date, put in to accommodate a new gateway or to replace the original without adequate cleansing of the ditch beforehand.

The spoil resulting from these operations is a problem and can become a source of much trouble and annoyance. It is often wet and heavy and, if piled up too close to the edge of the ditch, it soon begins to roll in again, while its weight and wetness may rot the ground beneath and give rise to serious falls of earth from the sides. Care should be taken to leave an adequate berm or margin between the foot of the spoil bank and the edge of the ditch, in order to prevent this. For the same reason it is desirable to spread the spoil as soon as it is feasible. If this cannot be done for some time, and especially on heavy land, it is worth while leaving or cutting gaps in the spoil bank to facilitate the get-away of surface drainage from the field.

The caving-in of the sides of new work is both aggravating and discouraging; it is moreover difficult to rectify. The causes are varied. A too steep gradient to the floor of the ditch may result in erosion and the gradual undercutting of the sides, which presently collapse. Insufficient batter to the sides in an unstable soil increases the risk of subsidence. The erection of a post and rail or wire fence too near the ditch may start falls near the posts. A heavy heap of spoil close to the edge, or traffic along the berm, encourages this tendency. Heavy rains, or frost and thaw, coming soon after the re-conditioning of a ditch, render its sides more liable to collapse. But perhaps the most irritating cases are those which occur as continuous or persistent 'running sand' trouble during the actual work. This is most liable to happen if the ditch bottom is in loose sand and the land alongside the ditch has a high water table and so is charged up with water tending to flow or drift towards the ditch. The sand 'runs' through lack of any cohesion, as a result of the water in it separating the individual sand particles from one another. The trouble is not easy to overcome. Heroic measures of revetting may be attempted, but these are expensive, and if they fail the ditch will be in worse case than before. The most hopeful line of action is to defer any attempted remedy until the water in the ground is at its lowest or the springs are at their weakest; to deepen the ditch gradually in shallow stages, especially if seepage ensues; to give a much flatter slope to the sides; and if the size of the job permits, to attempt stabilisation by turfing the sides. It may be possible in isolated cases to pipe the troublesome places, after a similar procedure, laying the tiles in material such as peat, straw, gravel, clinker, or turf, so as to keep the sand well away from the tiles, and minimise the risk of them silting up later.

The piping and filling-up of ditches, however, is not a practice to be lightly entered on or even commended. It is easy to point to examples of widespread harm which have followed the elimination of ditches. On the face of it there is no insuperable objection to replacing a ditch by a large tiled main; indeed, a number of advantages can be claimed, such as getting rid of hedges and their weeds, gaining valuable cropping area, enlarging fields to a size more in keeping with modern mechanised methods of farming, and avoiding the necessity for maintaining the ditch. But piped ditches themselves need care and maintenance, and there is no doubt that with the piping and filling in of ditches it is very soon a case of 'out of sight, out of mind'. Fortunately, amongst farmers and landowners there is no great enthusiasm for the practice, the more so since many of them have had ample opportunity to see the results of the elimination of ditches by individuals or authorities whose interest in and knowledge of the subject of field drainage has left much to be desired. Ribbon development provides many examples of it; road widening and straightening is responsible for others; and much of our urban development has necessitated the wholesale piping of ditches. Of late years, too, the rapid multiplication of landing grounds has involved this practice on a large scale. If it becomes desirable to pipe and fill in a ditch, certain important points must receive attention if trouble is to be avoided. They are:

1. The ditch must be thoroughly bottomed out and graded.

2. The pipe must be large enough to replace the ditch effectively.

3. It must allow drainage water to enter it, but not soil or silt in quantity.

4. All field drains which formerly ran into the ditch must be properly eyed in or connected with the pipe which replaces it.

5. These junctions should be recorded or marked, as they will no longer be visible or easily found.

CHAPTER IX

TILE DRAINING

For a decade at least, before the outbreak of war, the cost of tile draining had been generally regarded in this country as prohibitive. Ninety years ago, the cost varied from £2 to £5 per acre; just prior to the war of 1914–18, £7 to £10 per acre was the order of things; by 1938, £15 to £25 per acre seemed to be the prevalent idea. But by this date the amount of intensive tile draining which was being undertaken was negligible, so it is possible that the subject was being viewed in the wrong light. A thorough piece of work, involving surveying, planning, laying out and execution, to cover the whole area of a field with tile drains at definite depths and distance apart, complete with good outfalls, might well reach the high figures quoted, but nowadays such work is the exception rather than the rule. Even so, the costs might often be less. Two examples may be quoted. The first is one encountered in 1932 in circumstances of which the main features were flat land, on the Fen margin, with a ditch all round the field.

With 3 in. tiles, at 3 ft. depth and 1 chain apart, at 90*s*. per 1,000, and labour charges at 5*s*. per chain, the cost ran about £8 per acre. The second was on very stiff clay land in Cambridgeshire in 1938. In this case the depth was 20–24 in. and distance 12 yd.; there was a main and three leads with outfalls. With 2½ in. tiles at 70*s*. per 1,000 and labour charges on the minors of 4*s*. per chain (excluding bushing), the cost was of the order of £9 per acre.

At the present time, however, costs are higher in every direction, in labour and tiles in particular, so that schemes such as the above would nowadays cost a very great deal more. How do these figures compare with present-day events? Between July 1940, when the 50 per cent grant for tile draining was initiated, and the following October, schemes were approved for 5,600 acres of land, in all, involving an expenditure of £29,000. This can only mean that the average cost of tile draining in these circumstances was some £5. 4*s*. per acre. The acreage involved in these figures is the area receiving benefit from the operations carried out. It is necessary to seek an explanation of the apparently low cost of this tile draining, and the most probable reason is that most of the work which is being done is in the nature of repairs to or renovations of existing systems of tile drains. Some of the troubles that beset tile drains have already been indicated, and it is obvious that cases occur which can be rectified at a cost which, spread over the area of land which benefits, is remarkably low.

Tile drainage is commonly regarded as a permanent improvement. What the word 'permanent' means in this connection is not quite clear. Where they have been consistently cared for and maintained, there is no doubt that tile drains in light and loamy soils, with few exceptions, have functioned satisfactorily for 100 years or more. If neglected, however, they may soon come to grief and their benefit be lost (see p. 3 for data on land in need of draining). The case of really heavy land and the major clay areas is not the same, and there is room for doubt as to whether permanent improvement in the same sense can be effected.

It must be remembered that even before the rapid spread of tile draining in the middle of last century, under-draining had long been practised. The essence of the methods employed was the digging by hand of a drainage channel and the partial filling of it with any convenient coarse and durable material which was to hand locally, before the soil was replaced. Stones, bushes, horns have all been employed. Bush draining, of course, is still by no means uncommon. In Norfolk, old stone drains are of frequent occurrence, the stones for the most part being flints or pebbles. In Cambridgeshire one occasionally encounters drains made of chalk rock, broken lumps of those harder layers found at the base of the three main geological divisions of the Chalk formation, and much favoured in former times, where they were found near the surface, for flooring yards and steadings, making roadways and even for building purposes. In parts of the country where solid rock is found, the earlier under-drains were more substantial affairs and in truth were hollow drains, more or less square in section, with blocks of stone forming the side walls and the roof. Even granite blocks have been used for the purpose, where this is the local stone.

This being the case, it is not surprising that the principles and common systems of field draining were matters of importance and of considerable discussion before the days of drain tiles. Broadly speaking, there are three systems of draining, so far as lay-out is concerned, based on different principles. One of them copies nature. In any parcel of ground, given the opportunity, surface run-off would proceed in certain directions, decided by the shape of the ground surface. On a smaller scale it would resemble a typical river system, the main stream being found in the bottom of the valley, with its tributaries coming in from side valleys, where

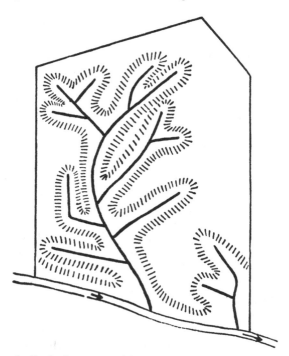

Fig. 17. A tile-drain system with a natural lay-out, i.e. occupying the valleys or 'lows'.

they collect the smaller rills coming down the slopes into them. Some field-drain systems are designed on similar principles, by siting the tile lines where the water, if flowing over the surface, would naturally go. They lead the water into the low places, then through them into the ditch, but all the way underground. Such a system is outlined in Fig. 17.

Another method, and this is probably the one which has been most widely followed, is to treat the wet area, or the whole field, as a unit to be drained, and to place drains uniformly over the whole of it with the object of removing surplus water wherever and whenever it may appear. Accordingly, the drains are laid at a set depth so far as is consistent with surface irregularities, and at equal intervals, to lead into a main. The depth and distance are decided by the depth and permeability of the soil. Such systems may assume a variety of shapes (see Fig. 18), but they all have one feature in common; that is, the whole area is evenly drained, and the minors in each area are placed parallel to one another.

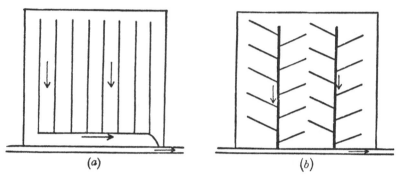

(a)   (b)

Fig. 18. (a) A 'thorough draining' lay-out with minors parallel to each other down the slope, and a main running parallel to the ditch, with a single outfall. (b) A similar field drained by two herring-bone systems with minors laid diagonally across the slope.

The third method is that which Joseph Elkington followed with such resounding success that the Government offered him £1,000 for his secret. From John Johnstone's account it would appear that the essence of it was to use all available evidence to deduce the origin, amount and direction of the movement of the water in the soil and then to place an adequate drain or drains to intercept it and prevent it from accumulating in particular places to their disadvantage. His line of thought obviously was 'Prevention is better than cure'. Of late years many odd bits of drain have been put in to 'cure' wet places in the field by providing a tile line to lead the water away from the wet place or to provide it with a way out, in much the same fashion as the average surface drain is dug during the winter. But as a well-known practical man often says: 'When you see water standing in a certain place in a field, it doesn't mean that it has rained there more than over the rest of

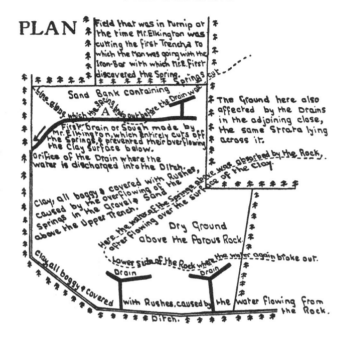

PLAN

Field that was in Turnip at the time Mr.Elkington was cutting the first Trench to which the Man was going with the Iron-Bar with which Mr.E.first discovered the Spring.

Springs cut

Sand Bank containing

Line along which the spring broke out before the Drain was A

The Ground here also affected by the Drains in the adjoining close, the same Strata lying across it.

First Drain or Sough made by Mr.Elkington, which entirely cuts off the Springs & prevented their overflowing the Clay surface below.

Orifices of the Drain where the water is discharged into the Ditch.

it was observed by the Rock, face of the Clay.

Clay all boggy & covered with Rushes, caused by the overflowing of the Springs above the Gravels Sand Springs in the Upper Trench.

Here the water of the Springs above the surface after flowing over the surface

Dry Ground above the Porous Rock

Lower side of the Rock where the water again broke out.

Drain   Drain

Clay all boggy & covered with Rushes, caused by the water flowing from the Rock.

Ditch.

SECTION

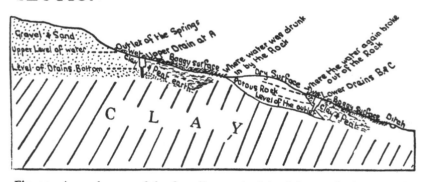

Gravel & Sand
Upper Level of water
Level of Drains Bottom

Outlet of the Springs
Upper Drain at A
Wet Clay
Boggy surface
Where water was drunk in by the Rock
Peat earth
Porous Rock
Dry Surface
Level of the outlet
where the water again broke out of the Rock
Lower Drains B & C
Boggy Surface Ditch
Clay Peat

C   L   A   Y

Fig. 19. A rough copy of the first diagram in John Johnstone's *Account of the Mode of Draining Land, according to the system practised by Mr Joseph Elkington.*

'Plan of Long Harol Pits, part of the farm of Prince Thorp in Warwickshire. Being the field in which Mr Elkington first discovered his Mode of Draining, An. 1764.'

'*Explanation.* *A* in the Plan represents the place where the Clay pointed up on the surface above the line and below the bottom of the trench, the depth of the field. That water has got there from other parts of the field. The object of draining should be to prevent it from getting there, not just to remove it after it has got there.'

In John Johnstone's account of Elkington's work there are some seventeen large diagrams explaining his drainage lay-outs in considerable and delightful detail. Fig. 19 reproduces the essentials of the first of them and admirably illustrates the 'interceptor' principle.

Though the accuracy of the geological formations in the accompanying cross-sections may be questioned, they well illustrate the vital facts connected with wet ground where any fall exists and the connection between the slope of the surface, the dip or slope of the strata, their permeability, the features of the water table and the resulting marshy conditions, is readily understood. The drains are sited towards the upper edge of the wet area, a point to be referred to later. It will be seen that correct diagnosis and one or two well-placed tile lines may achieve remarkable results with great economy of effort.

Before enlarging on the subject of lay-out, it may be desirable to draw attention to a few points in connection with the movement of underground water. It has already been explained that excess of rainfall percolates vertically downwards through the soil and subsoil so long as it meets with no obstacle. Any impermeable layer will check this movement and the water will pile up as a rising water table. There will be no considerable movement of this body of water unless the impermeable floor dips or slopes in a particular direction, in which case the water will move in that direction. In either case, the water table may come sufficiently near the surface to interfere with plant growth. Water standing on or very near the surface has obvious ill effects, but harm may arise through the water rising within 2 or 3 ft. of the surface for more than a few days on end. It is difficult to be precise on this point. It is well known that water-logging may actually be beneficial to the soil concerned if the water is fresh and moving through the

---

which, not reaching the spring, induced Mr Elkington to push down the Iron Bar, which at four feet below the bottom of the drain, caused the Water to burst up, and thus was the first means that led him to think of applying the Auger as a more proper Instrument in such cases, where the depth of the Drain does not reach that of the Spring, and upon this all his future practice has been grounded.'

soil, as in the case of water meadows; the position of the water table is not the only factor concerned. But in general it is considered that in arable land if the ground water persists within 3 ft. of the surface for more than a week on end, or in the case of grassland, within 2 ft., then the crops occupying the ground will suffer. The most familiar case is that of winter corn. Practical experience and controlled experiment alike have confirmed the importance of ground-water level in limiting the development of the root system of crops. Experiment, too, has shewn something of its effect at different stages of a crop's growth, in artificially controlled conditions. The actual position of the water table in the soil at any

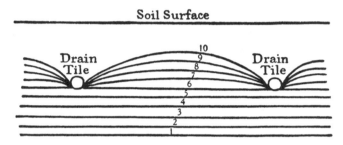

Fig. 20. 1–10, Successive positions of a rising water table in an open or free draining soil with a system of parallel tile drains.

moment, however, is the result of so many different factors that it is impossible to design drainage systems to give such rigid control of it as would seem desirable.

Considering the simplest case of regularly spaced drains at a certain depth in an open or permeable soil, in a level stratum, it will be seen that once percolation has begun, the ground water rises steadily with a level surface until it reaches the floor of the drains, when it will begin to enter them and run freely away to the outfall. Continued percolation will cause the water table between the pipe lines to rise to a higher level because of the appreciable time which it takes for the water to move laterally to the tiles from any point between them. The water table will pile up between the pipe lines, as shewn in Fig. 20, until equilibrium is reached and the pressure gradient becomes such that water enters the tiles as fast as it reaches the water table. The shape of this water table has a mathematical connection with the depth and distance apart of the drains, the permeability of the soil and the amount and incidence of rain. If it rains hard enough the water table between the drains will reach the surface in spite of their action. In between showers, on the other hand, the arch of the water table will gradually sink as the drains continue to run, albeit at a diminishing rate.

Even in light and loamy soils, however, the majority of cases are complicated by differences in constitution and permeability with depth. Most subsoils are denser, heavier and less permeable than the soils above them. Further, when disturbed by digging, as in the excavation of a trench for tile drains, the structure and permeability of the soil in the trench are radically altered. In most cases on re-filling, the soil is put back in a looser condition than it was before, except in so far as it is rammed or trampled in, a process which in a wet soil produces a poached or puddled condition, so that the filling becomes impermeable and water stands immediately above the tile line. Though this happens occasionally, the condition passes away in time and in any case it does not prevent the drains from benefiting the ground between them, especially in open soils. In the case of clay land, the material which is put back into the trench is much more open and permeable than before, a result of the many large interspaces which are introduced between the individual clods as they go back; this is made clear by the large surplus of earth which remains. If this fact is considered in conjunction with what has been said about the drainage properties of clay (see p. 28), it will be seen that the water table in clay soils, and so the functioning of the drains in them, is of a different nature. This is shewn in Fig. 21. The water table appears suddenly at the base of the topsoil and rises in it, the water in it flowing outwards to the drains by virtue of its small head of pressure, and percolating into them from above via the disturbed earth of the drain trench. The existence of free water and its movement is confined to the permeable topsoil and that in the trench.

The occurrence of slopes introduces another factor into the level and shape of the water table and gives rise to movement or drift of water underground in definite directions. There is a general tendency for the water to move in the direction of greatest slope and for it to accumulate in low places as standing water. It is most marked in the case of clay land on definite slopes and on those where there happens to be the outcrop of a clay formation overlaid by some more permeable stratum. It is in such circumstances that drains laid across the slopes or 'on the slosh' confer such marked benefit. The water moving down the slope encounters the drain and there finds an easier track to follow; this it does instead of continuing its slow progress down the hill parallel to the surface, with the consequent water-logging of all the ground until it finds a surface run-off lower down.

The practice of mole draining exploits to the utmost the peculiarities of clay land and the effect of slope. The extremely low cost of producing drainage channels by this mechanical means enables the operator to place his minors very close together, e.g. 8–9 ft., so that drainage water in the surface soil has nowhere any great

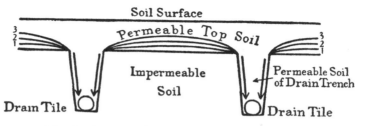

Fig. 21. 1–3, The development of a water table in clay soils and the movement of drainage water to tile drains via the disturbed and opened earth of the drain trench.

distance to travel to reach a drain and the way out. Furthermore, communication with the channel below is provided by the slit, left by the coulter which carries the mole or cartridge, and the fissures produced on either side of it by the actual passage of the plough through the ground. The moving of the ground for a distance of as much as 3 ft. on either side of the main slit is visible to the eye as the plough moves along, but it may not be generally realised that the subsoil in particular is fissured in a regular fashion, the cavities being vertical and opening into the main slit. This fissured portion of the subsoil facilitates the percolation of the surface water to the drain channels in the same way as the disturbed earth of a drain trench, but whereas that may occupy 18 in. in every 15–20 ft., in the case of mole draining it may be as much as 5 ft. in every 9 ft.

It is now possible to consider the different methods of draining in relation to the factors of water table and permeability. What we

called the natural lay-out or the placing of drains from individual
wet spots to the nearest suitable outfall, which, carried to its logical
conclusion, results in a system similar to a river and its tributaries,
is effective in getting rid of stagnant water by facilitating its
natural path. Persistently followed as occasion arises, this method
will undoubtedly achieve the desired end, but it is wasteful of time,
labour and material; it may entail costly errors, particularly as
regards the efficiency and capacity of the mains. The final result, too,
is an irregular network of drains, very difficult to record or to redis-
cover, in the absence of an accurate plan, when the occasion arises.

The second method, the essence of which is the laying of collecting drains at definite depths and regular intervals over the
whole area to be dealt with, is characterised as 'thorough draining'.
There seems to be some dispute as to the origin and meaning of
this term; in general 'thorough' is taken to be derived from
'furrow', the place where the drains were often sited. Be that as
it may, the present use and meaning of the word best describes the
attributes and objective of the practice. Whatever the variations
of soil, elevation of the ground, or the seasonal rainfall, such a
system should keep the field more or less uniformly free from
surplus water. If it does not, the probability is that the depth or
the distance apart of the minors is not the best for the local circumstances of soil and rainfall. In general, as is well known, the heavier
the soil, the closer and shallower the minors are placed. Fields on
uniform slopes lend themselves to grid systems; those which are
irregular, and possess one or more pronounced valleys, call for the
herring-bone lay-out from mains in the valleys; but in every case
the aim is to achieve success by a uniformly intensive system.
Whether some of the minors or parts of them are superfluous and
never run does not matter so long as the field is laid dry as a whole.
It is obvious that, given a set distance apart, there may be a
difference in the efficiency of minors laid down the slope and that
of minors laid across the slope. In deep open soils on very gentle
slopes there may be little to choose between the two from the point
of view of efficiency in lowering the water table, but there may be
point in adopting the former method to get enough fall in the
minors to give an easier run-off. On steeper slopes, however, the
general drift of the drainage water downhill may be so pronounced
that there is a tendency for it to continue moving downhill between
the drains, and so minors laid across the fall have a better chance

of intercepting it. There is added advantage, too, in not having water run down the minors too swiftly with a resultant check in velocity when it reaches the main. On heavier land, where the topsoil is often much more permeable than the subsoil and encourages a superficial drift of drainage water with the fall, minors across the slope are even more desirable. The simplest example of uniform drainage measures is found in association with flat land, with fairly permeable soil and ditches on all sides of the fields. Here it is frequently customary to carry the minors right through from one ditch to the other on the level. These minors are alike in every respect—diameter, depth and distance apart; and the water gets into the pipes and escapes from the two ends solely under the influence of hydrostatic pressure.

In the third method, that of Elkington, success depends on the correct diagnosis of conditions and on siting the drains in such a way as to achieve the desired result with a minimum of effort. Its merit lies wholly in its economy. On the other hand, the preliminary diagnosis may be wrong, and the drainage effort a failure. The most obvious example is the case of the spring line across a field, with most of the land on its lower side in a water-logged condition. It should be possible to lay a single drain in such a way as to divert all the water of the springs and lay the land quite dry. The same result, of course, might be obtained by a grid system or a herring-bone system down the slope so as to cover the whole of the wet area, but either of these would involve the use of much more labour and many more pipes.

It has been suggested earlier that the position with regard to tile draining in this country, except on clay land, is mainly one which calls for re-conditioning and repairs, rather than the laying of new tile-drainage systems. The digging out or thorough cleansing of ditches is working wonders on much of our light and medium land. But former neglect has, in some cases, been followed by breakdowns in the tile lines themselves, so that to dig out the ditches is not always enough in itself. If plans of all field-drainage systems existed, it would be an easy matter to check them up and see to what extent they were in order. But the importance of mapping and accurately recording all field-drainage work, while universally admitted, is frequently ignored. With the best will in the world, these most useful documents are apt with the passing of time to get mislaid, lost, or destroyed. Without a knowledge of the original dimensions it is not an easy matter to re-discover the lie of drains. Sometimes there is someone who saw or helped in the work from whom useful information can be obtained, but as often as not, little is known of all that lies hidden in the average field in the way of covered drains. Their position can sometimes be spotted during the inspection of a field, by means of certain indications already mentioned. Sometimes the slight ridge left by the excess of earth left on re-filling the trench may be visible; and after a few years this may be succeeded by a definite narrow depression, especially if much of the surplus earth was removed or spread about. The breakdown or stoppage of a covered field-drain system as often as not results in bursts which are fairly obvious, particularly during the drainage season. A stoppage will initially cause a wet area to appear, above and about it in flat or gently sloping land, or farther down the slope if the fall is pronounced. Rods, ordinarily used for clearing drains, can also be of great service in tracing the pipe lines. A common dodge is to push the rods along to their fullest extent, to note the general direction of the drain and to fix it more accurately by rattling the rods in the pipes while a second worker applies his ear to the ground. A hole is then dug to find and open the drain, when the same procedure can be repeated. Dyes such as permanganate of potash or fluorescine can be of service in identifying the outfalls connected with casually encountered drains, if they happen to be running.

Burst drains may be the result of a variety of causes, any one of which may so restrict the flow of the drain that water accumulates above the constriction until its increasing pressure causes the drain to burst at its weakest point. This may be at the seat of the trouble or it may be some way above it according to circumstances, which will decide where the water can force its way up to the surface and so escape. The channel originally may have been too small to deal with the amount of water attempting to run through it. The load depends as a rule on the area of land served, the type of soil and system, and whether water from outside areas also is concerned (see p. 58). Tile draining is a skilled occupation. Anyone can get the pipes in and covered, but it requires great skill to align them exactly and to get them laid on an even grade and on a firm bottom. These points are important, as faulty alignment diminishes the carrying capacity of the tile line, allows earth to be washed into the drain, and leads to possible further serious displacement of individual tiles on re-filling the trench. Weak, faulty, or damaged pipes are sources of trouble and should be discarded. Overground traffic of all kinds has in the past been the cause of damage to tile-drain systems, especially of recent years, during which neglect of drains has been so common. It is obvious that shallow systems are much more liable to damage in this way than are the deep ones. Timber hauling, tractors, steam tackle and the gyrotiller all claim their victims from time to time.

Silting is a problem in itself, though it may be localised in any particular case with its chief result a burst drain. It is generally due to the neglect of the ditch, leading to prolonged submergence of the outfalls; but it can happen, too, wherever there is a sudden diminution in the velocity of the flow of water in the pipe as a consequence of a sharp decrease in the amount of fall or of an increase in the diameter of the drain, e.g. at the junctions of minors with submains or of the latter with mains. Silting is of two kinds. In one it is primarily the fault of the soil; in the other it is due to bad draining work. It will be found in some cases that the so-called silt takes the form of a little heap at each joint between adjacent pipes. Such material is the accumulation of falls from above and about the gap between the pipes, over a period of years. The middle of the pipe in these cases is generally quite clear. The way to prevent it is to use uniform even tiles, laid close together, and to cover them first with a layer of turf, brushwood, straw, or other suitable material, before the earth is thrown back. The other type of silting, which results in the whole tile line or considerable lengths of it becoming filled practically to the top with fine mud, is due to the unstable nature of the soil concerned. The cause of this may lie either in its physical make-up or its chemical condition, particularly as regards its lime status. In either case it is the lack of structure or the unstable nature of the soil material which is responsible. The behaviour and properties of the surface soil constitute a criterion of some importance in the liability of the subsoil to cause silting. Where the crumb easily collapses and the soil runs together or pans under the influence of rain, the underlying material may have similar tendencies. In the case of sands of the finer sort, light silts, and clay sand mixtures, there may be little stability in the subsoil mass, and the condition known as running sand may be of common occurrence. This means that when the water table rises and drainage flow to the pipe line begins, the fine sand particles, having no cohesion with each other, are free to move with the water, and so they penetrate the joints and accumulate in the pipes wherever the current is not strong enough to keep them moving. But there are also a number of heavier soils, loams of various types, which lack stability. This weakness appears to be connected with the amount of free lime or chalk which is present in the soil. It has always been realised that part of the benefit derived from the old practice of chalking, in which large dressings of natural calcium carbonate were added to the land, lay in its effect on the physical properties of the soil. Cultivations became less laborious and the resulting tilth was more durable. In fact, chalking improved the soil structure and rendered it more stable. There are soils which are not so deficient in lime as to be markedly sour and so to need liming, for that reason, but which work better after liming, and evidence is accumulating that such soils are also prone to give rise to silting in the drains laid in them. In these cases, it is not the whole soil material which finds its way through the joints of the pipes, but the finest particles, i.e. the silt and, more particularly, the clay. These fine particles are not firmly held in the body of such subsoils, but under the influence of percolating water they break away from their neighbours and find their way amongst them to the drains. The drainage water in such cases is seen to be markedly turbid. It would appear that, at the end of each run of the drains, a thin layer of such material is deposited on the floor of the pipe where, on drying, it sets and adheres to it subsequently. In time the layer becomes thick enough to be obvious. If after each run the pipe line, through lack of fall in a particular section or because of periodical submergence of the outfall, is left filled with drainage water, a much thicker layer of silt will result and the choking of the drain is correspondingly more rapid. A uniform deposit, in which the layering of the material is easily verified, accumulates throughout the drain.

Another potent cause of trouble in tile drains are the root systems of trees or even of agricultural crops. Trees in the neighbourhood of mains and outfalls are particularly harmful. The roots, in their search for moisture, find and penetrate the tile line and develop rapidly inside it. It is not uncommon to find lengths of 15–20 yd. of pipe choked entirely with a matted mass of fibrous roots. In such cases the length affected must be relaid, and it is wise to do it with the help of glazed pipes and cement joints.

Timely attention on the first signs of breakdown in a field-drain system will often prevent considerable trouble later. The use of rods, or a stiff wire hauling a swab, with or without flushing, will deal with silting if this is not in an advanced stage. The process, of course, involves opening the drain every chain or so. Where it becomes necessary to re-lay short lengths of drain, as in repairing a break, particular care is necessary to re-establish the tiles on an even grade. To get such renewals on to a firm bottom, in addition, is equally important and even more difficult, but unless this is accomplished the new pipes will soon get out of alignment once more.

When a new drain or drainage system is contemplated, it is first necessary to decide on the general lay-out, which can be done in the light of factors already discussed (see p. 69). In the case of occasional drains siting is not such an easy matter and the success of the new drain may depend entirely on the skill with which it is placed. Such a solution of the problem is most likely to be sought in the case of localised wet areas situated on slopes and affecting only portions of a field. In some cases the water is thrown out at the surface by reason of the configuration of some impermeable stratum below ground. This often causes the wet area to appear first of all well up the slope and to spread gradually downwards, but sometimes it stops lower down, the result of more free percolation there. In such a case it should be the drainer's aim to place his collecting drain or his interceptor so that it can be got in deep enough to lie on or below the surface of the impermeable layer causing the trouble. This requires a deep drain across the slope within the wet area but towards its higher side, depending on how far the influence of the drain is likely to extend. If circumstances are such as to enable the drain to be so sited, a single line may thus cure and protect the whole area, but if the drain cannot be got deep enough to do this, some of the water may still pass below it to continue to give trouble lower down the slope, and so necessitate a similar drain parallel to the first but farther down. The effect of the first drain, if observed for a while, will help to indicate the need for and the best position for others to complete the cure. In this way springs can be tapped and dealt with much more successfully than by working on the idea of seeking a particular point where the spring is trying to issue and putting in a drain from that point. A line of springs is much commoner than a single point spring.

In some cases, however, the wet area appears first at the foot of the slope and spreads upwards, indicating a substantial or deep water table rising from below. Here there is reason in tackling the job from the bottom upwards. The interceptor principle may be invoked in both cases, but while the former may be dealt with by one well-placed pipe line, the latter may need something more like a thorough system.

In considering the approach to the problems connected with the details of new drainage systems, one cannot do better than quote from 'A Review of the Progress of Agricultural Knowledge during the last Eight Years', by Ph. Pusey, in the *Journal of the Royal Agricultural Society of England*, 1850:

Under-drainage...received a new impetus from the late Mr Smith, who coupled with it subsoil ploughing....Up to this time the depth of drains was under three feet. It is to Mr Parkes we owe the improvement of sinking them to four feet or five, and it does not detract from his merit if, as I believe, his rule upon the strongest clays suffers exception. I shall not enter on that debatable ground, but it is equally certain that shallow drains have been taken up, to be replaced by deep drains, and deep drains, in other places, been superseded by shallower ones. Draining, at whatever depth, for some years known to be profitable, is now indispensable, being only checked by want of means....The fact is, that we are too systematic in draining, especially when the work is begun upon a grand scale. The source of economy now must be in the maxim that 'one drain well laid to suit the circumstances will often save a dozen by rule'....Unfold your plan as you proceed....Again it has been said of late that drains should always go straight up and down hill. This is true, I believe, where the water to be drawn off is rain-water; but constantly a line of wetness may be found on a hillside, where the springs are thrown out, oozing through the field below. Draw your drain deeply along this line, and you will require no furrow drains lower down....I say then, that if, avoiding system in draining, you cut out your work to the substance of your land and its slope, draining should hardly ever exceed 3*l.* an acre all round....

The depth and distance apart of drains are closely connected in principle and many efforts have been made to give mathematical exactitude to the connection. They have not met with any marked success as a means of settling drainage problems in the field, because there are other uncontrollable and erratic variables which influence the final result. Given flat ground, a level water table, a soil uniformly permeable from surface to depth, and rain falling

at a steady known rate, it is possible to work out the depth and
distance of the minors in a drainage system which will keep the
water table from rising above a certain height midway between
the drains. Actually, however, the factors involved are not few,
simple and steady. Of the conditions named, flat ground is fairly
common and a level water table is found in some areas; but deep
uniformly permeable soils are very rare; while the irregularities of
the incidence of rain need no description. Thus it happens that
depth and distance are based on local circumstances, local experience and rough standards. In open free-draining soils the tile
lines are placed deep and wide; in impervious clays they are rather
put shallow and close. Even the meaning of the terms 'deep' and
'shallow' varies with locality, and with time. The 4 ft. minimum
associated with Government drains of earlier days would be distinctly deep to some present-day practitioners, yet there are circumstances in which deep draining at 6 ft. is still carried out. Probably
most tile draining to-day is laid within 3 ft. of the surface, except
near the outfall. Farming economics appears to have more influence
on it than logical procedure based on scientific facts. Even in mole
draining, the more expensive steam tackle work at a depth of
24–30 in. has given place to the cheaper practice of direct haulage
by track-laying tractors working at depths of 18–24 in.

There is an old-established rule which connects depth with
distance thus: a drain will affect the soil on either side of it to a
distance equal in open soils to 5 to 6 times its depth; in medium
soils 4 to 5 times; in clay soils 2 to 3 times. Drain intervals on this
basis would be as follows:

Depth ft.	Open soils yd.	Medium soils yd.	Clay soils yd.
6	20–24	16–20	8–12
5	17–20	13–17	7–10
4	13–16	11–13	5–8
3	10–12	8–10	4–6
2	7–8	5–7	3–4
1½	5–6	4–5	2–3

Such a range fits actual practice in only a few instances. Nowadays minors are rarely laid more than 3 ft. deep, or 3½ ft. at the
most. Such depths are common in the open soils at the edges of
the Fens; but the interval associated with them is 20 yd., not 12.
Drains at a depth of 6 ft. are to be found and are still laid in some

of the lighter Fen silts, but at 40 yd. intervals. Needless to say, deep drains of this kind are not the practice in clay land and it is rare to find minors deeper than 30 in. in such soils; even so, the intervals are larger than are shewn in this table. It does, however, agree fairly well with mole-draining practice, oddly enough. It will be appreciated that while in the case of open or permeable soils, especially where these are mainly composed of sand, there may be a comparatively simple relationship between depth and distance apart, the application of the principle breaks down wherever the subsoil is markedly less permeable than the topsoil. It has already been explained how the movement of the water to the drains is affected by such circumstances. The more impervious the subsoil is, the less likely is any benefit to accrue from deeper drains. In the light silt soils of the Fens, where the advocates of deep draining are quite convinced of the soundness of the practice, the moisture-holding power of the soil is too great for there to be any risk of over-draining, but the permeability is such as to allow steady if slow percolation. It is claimed that the deep drains run longer and later in the season, are less liable to silting, and keep the land drier. It is held that drains 6 ft. deep and 40 yd. apart are better than drains 3 ft. deep and 20 yd. apart, that they cost no more, and last longer.

If the sequence of events in the two types of drains in the same land is considered, the probability is that the water table sinks farther where the drains are deep than where they are shallow, the arch of the water table between drains becoming flat; this will happen during the more prolonged run of the deeper drains in early summer, the result chiefly of the greater distance between them. When the autumn re-moistening occurs, it may take percolation longer to reach the water table and cause it to rise. But while the deep drains will begin to run when the water table reaches them at 6 ft., the drains at 3 ft. will not commence in earnest until the water table reaches 3 ft. As the winter rains succeed one another, the arch of the water table between the drains rises, but in the rainless intervals it falls as a result of the action of the drains. It will be seen that with shallow drains it fluctuates about a much higher level than where the drains are deeper. This fluctuation within 3 ft. of the surface obviously may have an adverse influence on the productivity and earliness of intensively farmed land, as compared with similar deep-drained land. The depth and nature of the ditches in this type of country (the Fen light silt areas) confirms the importance attributed to deep draining. Such land is more or less unique, but the extent to which deep draining has been favoured in the past in other areas also is being shewn by the discoveries following thorough renovation of ditches.

Deep draining in clay or even moderately heavy land is more difficult to justify; indeed, the facts concerning the moisture properties of clay, and the movement of water in it, support the view that drains at 24–30 in. are the best in the circumstances. The whole problem is one of top water, i.e. water accumulating and moving in the topsoil. It cannot percolate to the drains except through the disturbed earth of the drain trench, so there seems little point in having this deeper than is necessary to preserve it from harm by weather, traffic or deep cultivations. Minors at 24–30 in. depth and 8–10 yd. intervals would represent typical tile draining in clay land. Fortunately, the mole plough has solved this particular problem with pronounced success.

Drain Pipes. From time to time efforts to solve the problem of field drainage by modifications in the form of drain pipes are to be noticed. In the early days of tile draining a big variety of shapes was tried out, but it soon became clear that the best general purpose type was the plain cylindrical form without frills of any sort. It is still the cheapest to make, the most convenient to cart, handle and lay accurately, and gives the best all-round performance, as regards durability and carrying capacity. The function of drain pipes is to provide and maintain an easy passage for water. The water gets to the drain as a result of the operation of the force of gravity; it flows there, and no property of the drain tile can accelerate this flow. The capacity of a tile line to clear drainage water is decided by its diameter and its fall; the main difficulty is to get the water to the pipe, not into it. With open joints, the space available for its entry is enormous compared with the area of cross-section of the inside of the pipe, which is all the space available for its escape at the outfall. The costs of draining being what they are, the only modification in tiles which would be attractive to farmers would be one which would make them simpler and cheaper; it is unlikely that anything simpler can be produced. This is not to say that there is no room for new ideas in drain tiles. The problems of silting, root trouble and making joints all offer scope for special forms of tile. For ordinary work, cheapness, uniformity,

Plate V. Thorough draining with tiles. (*See p.* 67.)

Plate VI *a*.   A 4 in. main in position for a mole-drain system, before bushing and re-filling. (By permission of *The Dairy Farmer*.)   (*See p.* 93.)

Plate VI *b*. The junction of a mole channel with the tiled main. The channel is visible in the side of the trench, as is also the slit gaping at the surface. The top of the tile can be seen below the bottom of the mole channel, and a piece of tile covers the joint below the junction. (By permission of *The Dairy Farmer*.)   (*See p.* 93.)

regularity of shape, freedom from flaws, and thorough baking should be looked for in the tiles used. Clay tiles should be straight-sided, square-cut at the ends, circular in section, hard, not too porous, neat, clean, free from blisters or cracks, and they should travel well. Glazed pipes, too, have their uses in spite of their greater length and the collar which they bear at one end. Seconds of this type can be used for making drains. With their joints unsealed, the fact that they are glazed is not a great drawback, but they are less easy to handle. Concrete pipes have not been used to any great extent for agricultural field drainage in this country; as competitors to the ordinary clay tile they are handicapped by their cost. There is little to find fault with in the simple clay tile, and until a cheaper competitor appears its use for agricultural purposes is unlikely to be seriously affected. According to American authorities there is little to choose between clay and concrete pipes in the matter of their general efficiency. There are much greater variations within each type than there are between them.

Size of Pipes. This point is generally discussed in terms of carrying capacity. The rate of free flow through a pipe under the influence of gravity depends on its effective internal diameter and its fall. The load the pipe may be called on to deal with is often considered in terms of the incidence of rain upon the surface of the soil. The rainfall over most of this country is 25–40 in. per annum. Here a fall of 1 in. in 24 hours is not a frequent occurrence; 1 in. within an hour is a rare event; $1\frac{1}{2}$ in. in 5 minutes is a historic event. It happened at Preston on August 10th, 1893. Now 1 in. of rain in an hour means that 22,400 gallons of water falls on an acre in that time; yet the maximum recorded rate of flow of agricultural field drains in open permeable soils in this country is 500 gallons per acre per hour; in heavy clay land, strangely enough, it is 1,500 gallons per acre per hour. Fortunately for farmers, however, there are two other factors which influence the amount and rate of run-off. These low figures are not due to any shortcomings in the drains themselves, or to inability of the water to enter them, but to the slowness with which the water moves through the soil to them. Soil, as we have seen, has the power to absorb considerable quantities of rain, a power which it steadily recovers between the showers. Roughly speaking, the total pro-portion of the individual falls of rain which runs away via the drains rises from nothing in late summer to 100 per cent on occasions in January and February, after which it falls off with the advance of spring and summer. It also takes an appreciable time for percolating water to reach the water table and for the surplus to travel to the drains from the part between the tile lines.

In actual practice, considerations other than those outlined are important in deciding on the size of tile to use. Those of small diameter are more difficult to align, more easily disturbed and much more easily blocked by falls or by silting, and these weaknesses are serious in tiles less than $2\frac{1}{2}$ in. in diameter. In his day, the famous Parkes advocated the use of 1 in. tiles, and his critics dubbed them pencil-cases. It is advisable on all counts to eschew the use of small calibre tiles. (See p. 132 for data on carrying capacity.)

Fall. There is sometimes a disposition to set rigid standards on this point. It should be borne in mind that much successful drainage work is carried out perforce on the flat. Fall is not essential to the functioning of a drain. But in the majority of agricultural land it is inevitable and, of course, desirable. At falls of less than 1 in 200, a meticulously careful and even grading is essential, to avoid dips in the tile line and consequent early trouble at such points. Even grading and a judicious inter-relation between the fall in the minors and that in the mains helps greatly towards the efficient and trouble-free performance of a drain system, particularly in avoiding silting, local hold-up of water, and bursts.

The Season for Draining Work. In practice, the time when draining work is carried out is decided mainly by convenience, particularly with relation to cropping and labour supply. From the point of view of the quality of the work done the best time is the end of the drainage season, when the soil is still soft for digging but is not charged with water, a condition which may result in the puddling of the trench bottom and instability of the tile lines. Where running sand is a menace, it is most desirable that work should be deferred until the water table is well down. Only at such times can a firm bottom be obtained and even then special precautions are worth taking. The use of long boards or a layer of topsoil or turf on which to lay the tiles, the wrapping of the joints in sacking, careful packing of the pipes with straw, hay, gravel, clinker, bushes or topsoil to keep the sand away from direct contact with the pipes are all aids to overcome this difficulty. Bushing-in or some similar procedure, even if it is only paring off a layer of topsoil from the sides of the trench direct on to the tiles, is sound practice, in order to lessen the risk of silting and the packing down of subsoil in direct contact with the tiles.

Outfalls. Gone are the days of elaborate Victorian outfalls, monuments to the solidity and thoroughness of the drainage work of those times. The principles on which they were provided, however, remain unaltered. The outfall of a drain system is a weak point and a vital point; the bank of the ditch near it is weakened if no precautions are taken. It caves in; erosion, frost, vermin and the treading of man himself conspire to weaken it; and once the rot starts, it soon works back along the drain to its detriment. Stability of the pipe line can be assured in varying degree without much expenditure by the use of long scrap-metal pipes or long glazed tiles with cement joints, anchored in small concrete plinths based on the undisturbed subsoil by the side of the ditch.

CHAPTER X

MOLE DRAINING

The following papers are reprinted with the permission of the several authorities for whom they were originally written by the present author; they deal with the subject of mole draining from different points of view.

THE TYPE OF SOILS SUITABLE FOR MOLE DRAINING

Where Mole Draining is Practised. The principle of mole draining has been on record for more than 200 years and it has been widely practised by means of a special plough since the latter part of the eighteenth century in the eastern counties—Suffolk, Essex, Herts., Beds., Hunts. and Cambs. Originally it was confined to those areas occupied by the outcrops of the Boulder, London, Gault, Kimeridge, Ampthill and Oxford Clays, as shewn in the Drift maps of the Geological Survey.

By far the greater part of the land which is acutely in need of field drainage to-day is of the type to which mole draining is suited, that is, the heavy or clay soils with very clayey subsoils. These, for the most part, are to be found in the areas named above, together with the clay districts of the Lower Lias, parts of the Keuper Marl and of the Weald Clay, in other parts of the country. While the operation of mole draining has been carried out in almost every county of England under the Government scheme of assistance, it has operated most intensively in the east of England. Of the 133,000 acres approved for grant up to the end of Sept. 1941, 112,600 were located in the counties of Beds., Cambs., Essex, Herts., Hunts., Leics., Northants., East Suffolk and West Suffolk, a fact which confirms the description of the clay areas given in the first paragraph of this note.

These, however, are not the only types of land to which the operation of mole draining is applicable. From 1925 onwards, the Ministry of Agriculture carried out demonstrations of the method at a number of centres widely scattered over the country. The work was performed successfully, and subsequent observation shewed that its results persisted for 3–9 years in areas other than those already mentioned, such as:

On Gault Clay in Bucks., Wilts. and Sussex; on Kimeridge Clay in Bucks., Dorset and Yorks.; on London Clay in Middlesex and Berks.; on Millstone Grit in Devon; on Oxford Clay in Bucks., Dorset and Wilts.; on Boulder Clay in Bucks., Lincs., Yorks., Lancs. and Northumberland; on Lias Clay in Warwick and Gloucester; on Keuper Marl in Cheshire; on Weald Clay in Surrey.

Drift Map as a Guide. In the eastern counties, especially, the Drift map (1 in. to the mile) has considerable value in assessing the possibilities of mole draining in a particular field, if used as explained in an earlier note. While a few fields and minor areas within the clay formations named may not be suited to the operation for other reasons, there are practically none outside these formations which have been really successfully mole drained; and as the boundaries of the formations can be transferred to the 6 in. O.S. maps with considerable accuracy, a valuable pointer can thus be obtained before inspecting the soil itself. The Drift maps can also be very helpful when fields which include the boundary between two formations are concerned. It is particularly important to know in such cases whether the lower part of the field is satisfactory or not, and the main difference between the two areas.

Examination of the Soil in the Field. It is in the field, however, that the final evidence is to be found. The character of the topsoil lies open for anyone to read, but this is not enough to rely on. If the soil is a real clay, hard and cloddy when dry, cracking markedly during drought, stiff and tenacious when moist, with much surface water-logging in late winter, its subsoil is almost certain to lend itself to successful mole draining. Not all heavy soils, however, are blessed with suitable subsoils for moling, and in the case of many heavy loams it depends whether there is a clay lying sufficiently near to the surface to make mole draining a feasible and beneficial proceeding. Hence the necessity for examining the soil to a depth of 2 ft.

Subsoil Characteristics. The colour of the subsoil reflects its drainage properties. Bright colours, especially shades of red, reddish brown, orange and yellow, if they are uniform or 'self-coloured', are the result of free percolation and good aeration.

The absence of aeration, or continuous water-logging, produces shades of blue, green and grey. A fluctuating water table eliminates self-colours and produces irregular, patchy or mottled colouring, i.e. flecks of orange, brown, or buff amongst the grey or blue background. The more open the soil in these circumstances, i.e. of a fluctuating water table, the brighter the mottling.

Accordingly, in assessing the suitability of a subsoil for moling, due regard should be paid to the following points. A smooth firm clay is suitable; a soft sandy clay is poor. A dull buff, or grey colour, with faint mottling, is a good sign; very bright and pronounced orange mottling is not. 'All top water and no bottom water' is associated with the real clays, 'top and bottom water together' accompany conditions less conducive to stability in mole channels, while 'all bottom water and no top water' means readily permeable and unstable subsoils. Amongst the more difficult cases are those of sandy clays, in which the clay contains enough sand to lessen its tenacity and give it a slight permeability, and the clay sands, in which there is just enough clay to destroy the marked permeability of the sand. Both are sticky, heavy materials through which water can move only very slowly; they take the mole readily but the channels produced are not stable and are liable to collapse more or less quickly. In this sort of medium, a big flush of drainage water early in the life of the mole drains may cause so much silting as to fill them entirely.

It is obvious that soils cover a whole range with regard to their suitability for moling, and this is reflected in the length of effective life of the resulting drains. Evidence is accumulating that the behaviour of a soil under this system can be related to the constitution of the subsoil on the following lines: where the subsoil contains more than 45 per cent of clay and less than 20 per cent of sand (as determined by conventional analytical methods), mole draining can be highly successful; where the subsoil contains less than 35 per cent of clay and more than 45 per cent of sand, it is unlikely to be of much use; where the subsoils have 35–45 per cent clay and 20–45 per cent sand, the length of life and the utility of the operation will vary accordingly, and can be enhanced by the judicious use of tiled mains.

Overcoming Weakness in the Soil. The kind and length of mains used in mole-drain systems can do a lot to overcome the weaknesses of the soil. While the use of moled drains can be justified in the more tenacious clays, it is unwise to rely on them in subsoils of weaker calibre. Tiled mains are well worth their cost in any circumstances, but with less stable subsoils they should be provided on a more generous scale, and instead of mole channels (minors) of length 200–250 yd. the allowance of tiled mains can with advantage be increased so that the maximum length of minor is reduced to 100 yd. or so.

As a general rule in heavy land the soil material gets heavier with depth from the surface and, in some cases, fairly loamy soils lie on stiff clay subsoils. This variation in texture with depth should be taken into consideration in deciding on the depth of the mole channels. It sometimes happens that moles drawn at 24 in. may have a reasonable chance of persisting and working, when at 15 in. in the same soil they would be so unstable as to collapse within a short time.

('Growmore' Leaflet No. 44)

MINISTRY OF AGRICULTURE AND FISHERIES

THE IMPROVEMENT OF LAND BY MOLE DRAINAGE

The Ministry of Agriculture is willing to contribute up to 50 per cent of the actual net cost of approved works of mole drainage, including piped outlets, the maximum contribution not to exceed £1 per acre (later raised to 30s.).

The scheme is administered by the County War Agricultural Executive Committees. In order to qualify for a grant, any scheme of mole draining must be submitted to the appropriate Executive Committee on a form which will be supplied by them. The Committee will examine the application, the work proposed and the estimated cost, and, if these are to their satisfaction, will recommend the appropriate grant. As soon as the Committee gives the necessary permission, work can be put in hand. Grants cannot be paid in respect of any work carried out before the Committee's permission is given.

The Value of Mole Draining. Poor drainage is one of the biggest obstacles to the successful farming of heavy clay land. Except in the top few inches of soil, such land is, practically speaking, impervious to water, with the result that surface water-logging is one of its natural features, with well-known adverse effects on the quality of grassland, the growth of crops (particularly winter corn), the ease of cultivations and stocking capacity. There is in this country a considerable area of such land that has been increasingly neglected as regards drainage during the past generation; at least 25 per cent of our heavy land is acutely in need of field drainage.

To drain such land adequately involves a very close or intensive system, and to achieve this by means of tiles is an expensive operation. Of late years, the cost has ranged from £10 to £20 per acre—a charge which the land in need of draining to-day is least able to bear. Mole draining is an entirely satisfactory alternative with many advantages over tiles in really heavy land, and one whose costs—15s. to 50s. per acre at the beginning of the war—are much the same as those of widely practised heavy tillages.

Mole Draining a Heavy Land Tillage. Mole draining is no new practice. It has long been the usual method of draining heavy land, and was much used during the latter half of the last century and the beginning of this as 'steam draining', with the help of steam tackle. Many of these 'sets' are still in work, but, since 1930, tractors also have been employed for mole draining.

The essence of the operation is to draw drainage channels in the subsoil clay by mechanical means. These channels persist because of the consistency of the clay; they are cheap to draw and can be placed close to each other, e.g. 9 ft. or even less. An important feature of the operation is that the mole plough has a regular opening or fissuring effect on the soil to the full depth of the channel, and on either side to as much as 2–3 ft. With channels 9 ft. apart, it will be seen that a good proportion of the ground is moved, and that mole draining is essentially draining combined with tillage. The full benefits are realised very quickly, and, although they begin to wear off after a few years, they may last much longer. Practical farmers on heavy clay land credit mole draining with an average life of 13 years.

The Position To-day. To keep heavy land up to the mark by mole drainage would involve the operation being carried out on about 7 per cent of the area of a farm each year. It has already been pointed out that at least 25 per cent of our heavy land is in need of such attention to-day. This figure is based on evidence examined in 1933. Since that date, such evidence as is forthcoming would indicate that not more than 12 per cent has been mole drained meanwhile. This is the state of affairs which the Government has in mind in offering to make grants to owners or occupiers of agricultural land in aid of approved works of mole drainage.

Ditches. The bad drainage of some land is largely due to neglect of the ditches. Frequently the thorough cleansing of ditches is what is required to put existing field drains into action once more, with substantial benefit to the land which they serve. On heavy land, especially where falls are poor, ill-kept ditches are a hindrance to efficient mole drainage also.

They often prevent the main and outfall from being given the necessary fall to ensure continuous clearance of drainage water, and in times of copious run-off cause the outfalls to be submerged, the mains to fill up and the whole system to silt up in its lower parts. Even the mole channels themselves may fill with water in the neighbourhood of the mains, and, as a result, collapse earlier than they normally would. Accordingly, the Committees will pay particular attention to watercourses, and will require to be satisfied that these are adequate before recommending schemes of mole drainage for grant. Improvement of the watercourses can be grant aided to the extent of 50 per cent of the net cost, under either the Agriculture Act, 1937, or the Agriculture (Miscellaneous War Provisions) (No. 2) Act, 1940.

Suitable Land. Mole draining is only suited to heavy land with a clay subsoil fairly near the surface and free from stones, pebbles, gravel, or veins of sand, all of which cause the channel produced by the mole plough to be ragged and liable to collapse after a short time. The surface must have an appreciable fall—enough to enable a surface-water furrow to function—and should be even and free from local depressions, so that the channel may have a similar even course. Though mole draining is most widely practised in the eastern counties, suitable soil conditions may be found on most of the clay formations of this country, and the system has been successfully employed in most counties.

Equipment. Various forms of equipment are available, the power required depending on both the size of the channel and its depth. For the heaviest work with $3\frac{1}{2}$ in. channels, at a depth of 24–30 in., double-engined steam tackle (or its Diesel-engined successor) is used. For 2–3 in. channels, at a depth of 20–24 in., the mole plough can also be drawn direct by one of the larger sizes of track-laying tractor, or by a less powerful wheeled or track-laying tractor fitted with a winch attachment. With the latter the tractor is anchored while the plough is drawn, and, as with steam tackle, the work can often be done when the surface is too wet for direct haulage. Ordinary types of farm tractor can be used only for relatively light work, such as 2 in. channels at a depth of about 15 in.

Siting, Distance, Depth, Length. Much clay land has little fall. As mole channels run parallel to the actual surface of the ground, they are usually so drawn as to run down the maximum slope in order to avoid the occurrence in them of depressions due to local surface irregularities. Where falls are steep enough, and the shape and size of the field are convenient, moles drawn diagonally across the slope may have an advantage. For effective drainage, the channels must lie close together, and it is usual to draw them at intervals of 3 yd. in really heavy land, though 5 yd. may be effective where the raw clay does not lie close up to the surface. This applies to all land worked 'on the flat', whether arable or grass. If, however, the land is laid up in high-backed ridges, as occurs in much heavy grassland, the place for the mole drains is along the bottom of the furrows. As the ridges may be of considerable width, it is desirable to draw sets of three mole channels in each furrow, one in the bottom and one on each side about 3–4 ft. away from it. Mole drains on the ridge are of little or no use. Surplus water quickly and naturally finds its way from the ridge into the furrow, where the mole channels soon receive it.

Mole channels function equally well at any depth from 12 to 30 in. Shallow work, however, does not last as long as deep work. To last well, the channel should be drawn in the raw clay, which is often more than 12 in. from the surface. Further, the channels should lie deep enough to escape damage by heavy vehicles or deep cultivations. There is less objection to shallow work in grassland than in arable. As a rule, mole drains should be drawn 15 in. deep at least, and it is all to the good to have them deeper.

On the average, steam tackle works at a depth of 24 in., and the drains last for 15 years. Equivalent deep work by direct haulage demands the use of heavy track-laying tractors.

The choice thus lies between the more costly, more durable type of work and the cheaper type of operation which may need repetition after a shorter interval. Mole drains, with their narrow spacing, have a water-carrying capacity equal to most demands that are made on them, and the matter of their length is usually decided by the size of the field or the best sites for the collecting mains.

Mains and Outfalls. Much of the value of a system of mole drains depends on the siting of the mains and the provision of suitable outfalls. If the moles have been drawn where there is an adequate fall, the best method is to provide a well-laid tiled main, cutting across all the mole channels towards their lower ends, and leading

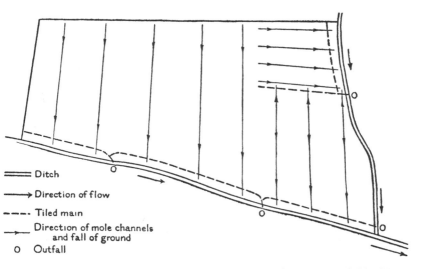

Fig. 22. Lay-out of mole channels and tiled mains in a 17 acre field with a satisfactory fall (more than 1 in 200).

into the ditch at a convenient place (see Fig. 22). For this, 4 in. tiles serve very well. They need to be laid deeper than the mole channels and covered with a layer of brushwood and clinker before the trench is filled in again to such a level that the mole channels can 'weep' into the main. With such a main, it is possible to repeat the mole draining quite easily, when necessary. A length of cast-iron water pipe, with its end covered with wire netting to exclude vermin, serves as an efficient outfall.

Where the fall is insufficient (less than 1 in 200) and irregular, namely, where the field as a whole is more or less level, a more effective lay-out may be obtained by siting several mains in such a way as to take advantage of all natural low places or valleys, and

drawing the moles so that their whole length is divided amongst
the mains according to the irregularities of the fall (see Fig. 23).

A cheaper, but much less permanent, form of main can be
provided by using the mole plough itself, with a larger expander
perhaps than is used for the minors, to draw several channels along
the desired track, deep enough for the minors to clear them when

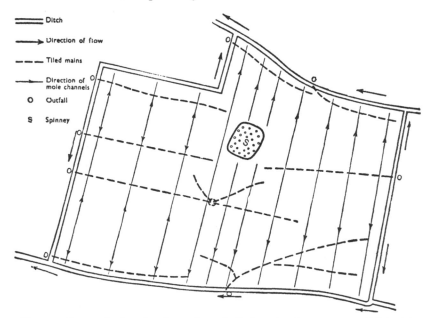

Fig. 23. Lay-out of mole channels and tiled mains in a 55 acre field with
irregular and inadequate falls (mostly 1 in 300 or less). For the most part, the
ground falls away from the spinney, occupying the highest part of the area,
to the surrounding ditch.

drawn. Connection may be improved by the use of the auger or
crowbar. If this method is employed, the mains should be drawn
first. It will still be necessary to provide outlets in the form of tiled
leads 15–20 yd. long to connect the moled mains to the ditch (see
Fig. 24).

Mole drains can be drawn sometimes from the ditch itself, but
each one must be finished off by inserting one or more 2 in. pipes
or a length of boiler tube as an outfall, or else the ends quickly
collapse and the channels cease to function.

The fall is so slight in much clay land that ditches do not afford adequate clearance for outfalls, and in addition are apt to silt up fairly quickly. To ensure that the mole-drain system gives its maximum length of service, it is important to keep the outfalls clear. Failing this, the water will back up the mains and hasten the decay of the channels at their lower ends.

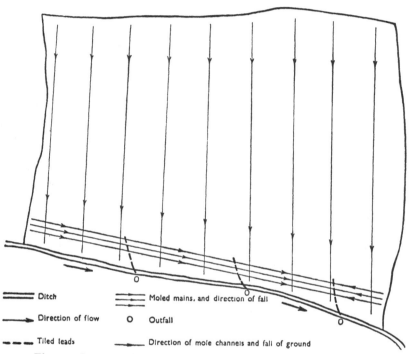

Fig. 24. Lay-out of mole channels, moled mains and tiled leads in a 21 acre field with satisfactory falls.

Costs. The many varying factors in mole-draining operations make it difficult to give a close estimate of the cost. It is certainly only a fraction of the cost of tile draining. Land has been mole drained and supplied with moled collecting mains for as little as 15*s.* per acre. On the other hand, mole draining with channels at 20 in. depth and 3 yd. apart, together with tiled mains and outfalls, has cost 50*s.*–60*s.* per acre. These instances indicate that costs depend upon individual circumstances. In schemes whose total cost is of the higher order, it is found that one-third is accounted for by the drawing of the moles, one-third by the digging, laying and refilling of the tiled mains, and one-third for materials (tiles and outfall pipes). By dispensing with mains the last two items can be eliminated; or by using moled mains with occasional tiled leads they can be reduced very considerably; but both economies are achieved with much loss of efficiency, and it is often impracticable to dispense with mains.

Mole-draining costs are of the same order as those of some of the cultivations carried out on heavy land. The benefits of mole drainage entitle it to be regarded as a cultivating operation that ought to be carried out more often. As a means of ridding clay land of its superfluous surface water, the merit of the system is firmly established. Further, the cost of tile draining is so high to-day that the possibilities of the mole plough on heavy classes of land generally are worthy of serious consideration. It is frequently noticed that mole draining revives the activity of pre-existing tile drains, a fact which in itself suggests that there are possibilities in a mole-drain system laid over skeleton tile drains, or deliberately undertaken for renovating old drainage systems in land not of the heaviest type.

Where Mole Draining is Most Needed. On any given heavy-land farm, even if the individual fields lie on the same soil formation, there is not at any one time the same need for mole drainage throughout. A long ley endows the soil with much improved drainage properties. Old grassland is even better in this respect. This benefit wears off after a few years in arable cultivation and disappears, and in heavy land which has 'fallen down' while in a state of low fertility, natural drainage properties are inconspicuous. At the present time, therefore, the greatest need for mole drainage and the most marked results will be found in old arable land and in semi-derelict land, a fact which should influence farmers in deciding which fields to do first.

General Considerations. Mole draining is best carried out when the subsoil clay is moist and plastic enough to allow the mole to pass through it smoothly; when, with direct haulage, the surface is dry enough to afford the tractor a firm grip. The best conditions occur as a rule between October and May. After a dry summer, however, the subsoil may not be fit until December; in a wet winter, there may be long periods during which the surface is not fit to carry a tractor.

Much mole draining is done in March, April and May, after the immediate surface has dried a little. Grassland and stubbles present no difficulty, and even land in winter corn may be mole drained without undue harm to the crop if weather conditions allow of the tilth layer drying sufficiently to enable it to take the tackle without damaging the topsoil condition.

If the plough is correctly set and the subsoil is in the right condition, the mole glides smoothly and a fairly smooth-walled channel results. If, however, the plough is out of adjustment or the subsoil is too dry, a distorted and rough-walled channel results, which will inevitably collapse more quickly. As the plough moves along, it can be seen that the ground on both sides of the coulter heaves, with heavy machines, for about 2 ft. on either side. This movement is accompanied by actual fissuring of the ground above the channel. The fissures are mostly vertical, but they join the main slit. The slit and fissures together provide the means by which the surplus water reaches the mole channel. Definite advantage is to be gained by drawing mole drains uphill rather than downhill. It has been demonstrated that drains drawn uphill clear the water at a greater rate and silt up less quickly than do drains drawn downhill.

The vertical slit made by the coulter persists for years, opening in times of drought, but becoming invisible in moist periods. Sometimes this gaping extends as deep as the mole channel or even beyond it, and harm may result from loose material falling into the channel, or through the walls closing in and obliterating it in places on re-moistening. It will be seen that there is a definite connection between depth and durability, deeper channels being less susceptible to the adverse effects of the weather.

A good mole drain system with moles drawn 20–24 in. deep and 9 ft. apart will give excellent results for 6–7 years and quite useful service for a further 6 or 7 years, in heavy clay land. Such a system is capable of being reconditioned simply by drawing the moles afresh over the original mains.

Points on Mole Draining[1]

Mole draining consists essentially in drawing a bullet- or spindle-shaped steel mole or cartridge through the subsoil at a given depth from the surface of the ground so as to produce a drainage channel. The cartridge is fixed to the lower end of a vertical blade which in its turn is mounted on a suitable form of carriage, which can be drawn along the desired track by mechanical haulage.

Suitable Soils. The operation gives the best results where the subsoil is uniform undisturbed clay to within 12 in. of the surface, and where there is a fall of the order of 1 in 200, or more. The surface of the ground needs to be free from minor irregularities and the clay subsoil without pockets of sand, gravel or pebbles. Wherever these occur, a mole channel, otherwise satisfactory, will be weak and less durable. When it collapses, which happens comparatively quickly, the rest of the channel above that point is put out of action. When a channel is drawn through pockets of sandy, gravelly, or pebbly clay, the walls, instead of being smooth and coherent, are torn and honeycombed, with the result that they soon collapse when drainage begins.

Effects of the Operation. It is important to realise what happens in the actual operation of mole draining, as certain features have a marked influence on the final result. It is impossible for the cartridge to start its work at the level at which it is desired to place the drain, unless it begins in the ditch or in an eye hole specially dug. The former is often impossible because of intervening hedges or fences, and in any case the entrance to the channel is ragged and loose and as a result it easily collapses and becomes choked. Outfall pipes can be inserted in each channel but, even so, are unstable and unsatisfactory. Eye holes, moreover, are laborious to make and need excessive manoeuvring on the part of the plough operator. Present-day mole ploughs will enter the ground and assume their proper position automatically. This is the course usually followed, a start being made at a point some yards from the ditch. As the plough moves along, it can be seen that the ground heaves on both sides of it to a distance of 1 ft. or 18 in.

[1] Revised and reprinted from *Husbandry*, the quarterly publication of the Norfolk County Council Department of Agricultural Education.

Plate VII. A main which has been bushed prior to re-filling.
(By permission of *The Dairy Farmer*.)   (*See p.* 93.)

Plate VIII *a*. A plaster cast of a mole drain drawn in smooth clay, and another of the same channel a few yards away where the clay contained a small amount of gravel. (By permission of *The Farmer's Weekly*.) (*See p.* 98.)

Plate VIII *b*. Two casts shewing the effect of wear of the mole plough or of faulty setting. These two channels were produced by mole ploughs of identical make and calibre. Note the distortion and tearing of the channel on the right, the result of the digging effect of a badly adjusted plough; it very quickly collapsed and disappeared. (*See p.* 100.)

The Slit and Fissures. The ground is obviously fractured and opened up in the wake of the plough, from the surface down to the channel itself. Investigation has shown that the slit made by the blade, and the accompanying fractures, persist for a long time. Whenever the ground dries out, particularly in summer, the slits gape to a marked extent, obvious in grassland to the casual glance, and equally obvious in arable land, if the tilth layer be removed by digging. The slits have been seen open in grassland, in a drought year, as much as 16 years after the mole drains were drawn. The fractures are not so obvious, but they have been examined by means of plaster casts taken shortly after the operation. They open out from the vertical central slit like so many vanes or flanges, and the extraordinary feature about them is that they all make an angle to the line of the mole, an angle whose apex points in the direction from which the mole plough travelled. This means that the slit carries a profusion of short branches making a typical herring-bone system.

Direction of Pull. The direction of pull, then, is important because, if uphill, the fissures facilitate the entry of drainage water into the channel more than if the mole is drawn downhill. In effect they conform to ordinary tile-drainage practice. This feature is of importance in the channel itself. The floor and sides tend to lift or peel behind the cartridge. This tendency is more marked if the subsoil is on the dry side. In any case the resulting serrations point backwards from the plough just as the accompanying fissures point towards its direction of travel. They are more easily broken off by running water if the channels are drawn downhill than if they are drawn uphill. Actual experiment has shown that moles drawn uphill run quicker and clear more water than moles drawn downhill and that the latter silt up more quickly than the former. But this fracturing of the surface of the channel is undesirable, whichever way the edges point, as it is a source of weakness and leads to an earlier collapse of the channel. When the cartridge leaves the ground it inevitably tears or bursts its way to the surface and the resulting disturbance is most marked. On many counts then, the drawing of moles uphill only, an old-established practice, is to be commended, and the natural inclination to draw up and down alternately should be resisted. Moles drawn uphill are distinctly better than those drawn downhill and the operation leaves the bottom of the field, where it will probably be desired to lay a tiled collecting main, or to draw two or three larger mole channels as collectors, free from points of disturbance or weakness in the minors.

Changes in the Channel. The internal condition of mole channels and events therein are, of course, not visible to the eye, but it has been possible to study them by taking casts at intervals. A mole channel is frequently depicted or imagined as of circular cross-section, but this is rarely the case. Two things prevent it, namely, the set of the cartridge, and the natural resilience of the clay under the pressure produced by the passage of the cartridge. The channel, as freshly produced, is ovoid or egg-shaped in cross-section, with the long axis vertical. Even a spherical expanding bob, attached to the cartridge, does not leave the channel round. It helps to enlarge the channel and to smooth the walls. For instance, in an example measured recently, a $2\frac{1}{2}$ in. cartridge, with a 3 in. bob attached to it, left a channel of ovoid shape $3\frac{1}{2}$ in. from roof to floor, and $2\frac{1}{4}$ in. from side to side. In the matter of smoothness of wall and the shape of the channel, the important factors are the moistness of the subsoil clay and the set of the machine. With age and wear the above-ground adjustments of the mole plough frequently work loose and assume an excessive amount of play. This results in the cartridge getting out of alignment and being dragged through the ground in a digging fashion. The vertical depth of the channel becomes grossly exaggerated, e.g. from 2 to $5\frac{1}{2}$ in., and the walls badly torn. Such a channel lasts a very short time.

The Right Season to Mole Drain. As indicated above, the moistness of the subsoil greatly influences the durability and quality of mole channels drawn in it. But in deciding the best time at which to operate, there are other considerations. The surface must be dry enough to carry the implements, whether they be steam tackle, wheeled or track-laying tractors, without excessive damage to the surface. Direct haulage by wheeled tractor, especially if there is no great margin of power for the job, is even more dependent on suitable surface conditions. The right conditions of surface and of subsoil are most likely to coincide in autumn and spring, and more certainly in the latter season when chances of the subsoil being moist are greater, even if the surface is dry. After the summer, especially when it has been a droughty one, the soil is moistened by autumn rains from the surface downwards and it may be the New Year before the subsoil recovers its moist condition, by which time the surface may be in a slippery and sticky state. A few trial holes with a spade will always reveal the condition of the subsoil clay. As a matter of fact, seasons vary considerably. In the winter of 1936–37, for instance, in most clay areas in the eastern counties, mole draining was out of the question from November to May, because of excessively wet surface conditions, but the operation was carried out very satisfactorily in June and July 1937, when the surface became firm. The autumn generally provides periods favourable to this class of work, when the clay has not dried out to great depths in the summer.

Depth. It is by no means easy to decide what is the best depth for mole drains. In some heavy land, a stiff clay subsoil is to be found immediately below the tilth layer, quite stiff and plastic enough for successful mole draining; but in many cases it is necessary to go deeper. In any case, it is advisable to draw the channel deep enough to be out of harm's way from traffic or cultivations overhead, and to be below the depth to which the weather, i.e. frost and heat, and drying-out, penetrates to a marked degree. Shallow moles are much more liable to decay and collapse due to seasonal weather changes than are those which are drawn deep. Inspection of the sides of a hole dug in clay land will shew how far the weather penetrates; but it will as a rule be found that the unweathered clay is encountered within 2 ft. of the surface. In practice, moles are drawn at depths varying from 12 to 30 in.; the most desirable depth is from 20 to 24 in. It will, of course, be realised that the deeper the channel wanted the larger and heavier the tackle necessary, and the more expensive the operation. While a 10 drawbar H.P. wheel tractor can draw a 2¼ in. cartridge at a depth of 16 in., present-day contractors with high-power track-laying tractors work comfortably with a 3 in. plug at a depth of 24 in.

How Mole Drains Work. The movement of water and the accumulation of excess water in heavy land take place in the tilth or surface layer, and drainage depends on giving the water access to the drainage channels. In mole drainage this is via the slit and the disturbed ground in its neighbourhood, so that it is doubtful whether deeper drains act more efficiently than shallow ones, except in so far as heavier and larger implements open up the ground to a greater distance on either side of the slit. In actual fact, deep channels come into action as a rule later than shallow ones, but they continue to run longer. The difference in time of starting to run is rarely more than 2 or 3 hours, and in the actual amount of run-off there appears to be little difference.

Distance. This being the case it is doubtful whether any connection between depth and distance can be established. Obviously, in a general way, the closer the drains are together the better will be the final result. Practice varies between 5 and 30 ft. In arable land, and where the ground is free from ridges and furrows, a distance of 9–12 ft. is desirable and effective. But where pronounced ridge and furrow exist, the channels should be drawn in the bottom of the furrow. Drains on the ridge are of little or no use. The water naturally flows towards the furrow and here is the place to site the mole channels, of which a trio in each furrow, at 3 ft. intervals, are sufficient to deal adequately with the run-off.

Length of Life of Mole Drains. The question is often asked: 'How long will mole drains last?' It is not very easy to give a satisfactory answer. This will readily be realised by a consideration of some of the points dealt with above. The factors which influence the life of a mole drain, accidents excluded, are the depth, the type of subsoil, the set of the mole plough, the moistness of the subsoil at the time of drawing, and the seasonal weather changes subsequently. In this connection, the results of an inquiry made by R. McG. Carslaw a few years ago over a large number of farms in Suffolk, Essex, Hunts. and West Cambs. are of interest. Practical opinion varied in putting the life of mole drains at 2 to 50 years. These limits are obviously extreme; at the former the soil was admittedly unsuitable; at the latter the author was an enthusiast. The majority opinion gave them a life of 10 to 15 years, and this in districts where mole draining has been carried out by means of steam tackle for many years. Most of this work has been done at depths of 20–24 in. Opinion varies with districts; apparently the nature of the subsoil is the deciding factor. Thus, in general, a longer life is obtained on the Boulder Clay of Suffolk and north-west Essex than on the London Clay of south Essex. There is also widespread agreement that mole drains last a year or two longer in grassland than in land under the plough.

From 1924 onwards, the Ministry of Agriculture carried out demonstrations of mole draining on many classes of land throughout the country. The work done on these occasions has been kept

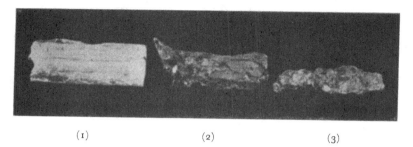

(1)   (2)   (3)

Plate IX *a*. Changes in mole channels with age. Casts shewing the condition of a mole channel (1) when freshly drawn, (2) after two years, and (3) after four years of life. (By permission of *The Farmer's Weekly*.) (*See p.* 104.)

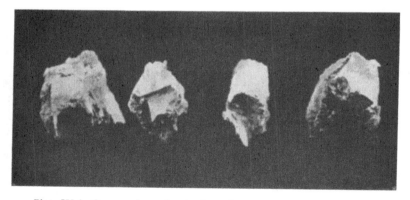

Plate IX *b*. Cross-sections of mole channels two years old. (*See p.* 104.)

a

b

Plate X *a* and *b*. The steam tackle, hauled by twin steam engines and steel cable.
(*See p.* 110.)

under observation, and J. H. Blackaby, reporting on the results, states that mole drains drawn by direct tractor haulage at depths of 14–16 in. have functioned satisfactorily for 5 to 9 years. This, for the major part, was shallower work and of less diameter than is done by steam tackle or by modern heavy track-laying tractors, and this fact may explain the difference in the effective life found in the two investigations. It must also be borne in mind that some of the work was still of service when reporting on it ceased.

As a matter of fact, mole drains act very efficiently in the first 2 or 3 years of their life and then steadily deteriorate. There is no doubt that in some cases they run for many years, but they may do this and still fail to clear the land of its excess of water sufficiently quickly. Thus it comes about that frequently it is profitable to repeat the operation after a life of 5 or 6 years. The writer has a number of mole drains under observation and running freely after 10 years' life, but water often stands on the surface above them and it was found advisable to re-mole the field as a whole in the sixth year.

Causes of Decay. It has become fairly obvious that the nature of the soil, and more particularly of the subsoil, has a fundamentally important influence on the length of life of a mole drain, and most observers seem to be of the opinion that unevenness of the subsoil, and especially the presence of sand and gravel in it, are most responsible for the decay of these drains in land which is otherwise suitable for the operation. But even with good uniformly stiff clay, conditions during the drawing of the moles and the subsequent weather have important effects. It is in this connection that depth of itself can play an important part. The fluctuations of heat and cold, dry and wet, obviously get smaller with depth below the surface, so that deep mole channels are much less subject to decay from these causes than are shallow ones. It is a matter of actual experience that on any one class of land, shallow work lasts a shorter time and needs to be repeated sooner than deep work. Against this may be set the smaller effort required and therefore the lower cost of shallow work.

Silting. The way in which clay swells or shrinks as it moistens or dries out is well known and is the cause of the cracking or fissuring of the ground during dry spells. Even in normal seasons, this affects mole channels, whose dimensions steadily decrease apart from any decay or collapse, as actual measurement over a period of years has shewn. The stresses and strains about the channel, however, are uneven, and bits begin to fall off the roof and drop on to the floor of the mole. They do not cause so much obstruction as might be expected because in a stable kind of clay they do not rapidly collapse into silt. It is probable, nevertheless, that 'silting' brought about in this way is much more serious in some soils than in others. Examination of the contents of 'silted' tile drains shews that in some cases the material is much the same as the surrounding soil; in others it is composed almost solely of the finest particles of the soil.

The two main factors influencing the tendency of a soil to silt up its drains are its physical make-up in terms of the amounts of sand, silt and clay which it contains and its lime status. Well-limed soils are much more stable in their structure and their resistance to weathering, while soils deficient in lime tend to collapse and disintegrate. In a comprehensive experiment on mole draining, carried out by Dr E. C. Childs and the writer, it has been observed that the drainage water from limed plots is much less turbid and carries much less solid matter than that from the same soil in an unlimed condition.

Effect of Drought. The effect of extremes of weather is also important. In summer drought the fissuring of clay soils proceeds to great depths, and it sometimes is possible to lower a rod 2 or even 3 ft. into a crack, i.e. to a depth beyond that of most mole drains. This is particularly the case in grassland. The cut made by the blade coulter of a mole plough leaves a plane of weakness which lasts for many years, and every time a drought occurs the slits gape open and sometimes are prolonged to a depth below the mole channel. The question at once arises as to whether this is a serious matter. Observation of mole-drained arable land leads to the opinion that surface tillages to some extent prevent this, but not entirely, for mole-drain lines are sometimes visible in a dry stubble. In any case, the masking effect of a tilth layer is confined to that layer, and, as digging has shewn, the slits during drought still gape open below it. In grassland, efforts have been made to prevent the opening of the slit, by ploughing out a furrow along which the mole is subsequently drawn, and later replacing the slice *in situ*, a policy of perfection which is of doubtful value. As a matter of fact, in grass or in arable land, a certain amount of soil undoubtedly falls down into the channel, but it is unimportant in comparison with another result. This is, that after the channel has gaped open, the subsequent swelling of the body of the clay about it, on re-moistening, may be so pronounced as to cause its walls to close in on one another to such an extent as to reduce it in places to a mere shadow of its former self.

The effects of drought, serious as they may be to the shape and size of the channel, are not wholly adverse. They are beneficial to the drainage properties of the soil as a whole, as this is rendered more open and more freely draining, so that even an impaired channel can function usefully to get rid of the percolating water. Inordinately wet seasons, especially of the type where two wet winters are separated by a moist summer, have a very bad effect in closing the natural cracks and fissures so that the passage of drainage water to the mole channels is much impeded and water-logging is much more persistent as a result. Thus it comes about that clay land fares better after a dry summer than after a wet one.

How Drains Run. No drainage comes from clay land until it has absorbed moisture up to its capacity, so far as the subsoil is concerned. How soon this happens depends on the dryness of the preceding summer, more than anything else. Now, in England, as a rule, the amount of rain falling on the land begins to exceed the amount which can evaporate from it, towards the end of August. From this time the soil begins to moisten up. The nature of clay is such that in one year, after a dry summer, the land may be able to absorb the equivalent of 11 in. of rain without the drains running. It even happened, in 1933–34, in many areas, that there was not enough rain during the winter to produce an excess of water in the soil, so that the drains never ran that season. On the other hand, after the moist summer of 1936 these same soils could only absorb 4 in. of rain before they became saturated and their drains began to run. This amount may come in a few weeks, in which case the drains (heavy land) will be running before September is out.

It is interesting to keep a close watch on the performance of drainage systems. As a rule, light-land drains will begin to run earlier than heavy-land drains on the same farm. Their behaviour is more akin to that of springs. When the water table in the land reaches their level, they begin to run and continue to do so for a long time. The fluctuations are not very violent and they will continue for days at a very gradually decreasing rate, after the rainfall which caused them to run. Heavy-land drains, on the other hand, are subject to much more rapid changes. Once they have started to run, their rates rise and fall with extraordinary rapidity. This is the more marked where the system is one of mole drains. Thus, in a particular case, on a very wet day when it was expected that some clay-land mole drains would begin to run, it was found that they were all still dry at 5 p.m. But at 9 p.m. the same evening, the grassland drains had started and one of them was running at 1,500 gal. per hour. At 11 p.m., only 2 hours later, the rate was down to 700 gal. per hour. It rose again during the night, at 9 a.m. it was down to 700 gal. per hour again, and by the evening, less than 24 hours after it had started, the rate was down to 100 gal. per hour. On similar clay, in neighbouring fields in arable cultivation, the flow did not commence until about 3 a.m., some hours after the grassland drains had begun. The rate reached a maximum of 70 gal. per hour at 2 p.m., and very slowly fell off during the next few days. These facts illustrate the caution necessary in forming an opinion of the efficiency of any drainage system, and shew how very different are the two cases of arable and grassland, even on the same soil type. The effect of cultivations is to make the drainage behaviour of clay soils tend more to that of open soils. That is, the flows are longer and more equable.

Advantages of Mole Draining. Mole drainage is sometimes criticised, especially by those who favour deep drainage, on the score that it only removes surface water from the land. But in the type of land to which mole draining is best suited and in which it works effectively, surface water is the only part of the water content of the land which can be got rid of by drainage. In clays, free percolating water does not exist except in the surface soil. The moisture below is not free to percolate even if drainage channels existed to carry it off. Once the surface water has been drained away, however, the beneficial influence of sun and wind can get to work to improve conditions in and below the topsoil. The old saying that 'though drained the land, 'tis clay land still', is as true to-day as ever it was, which explains to some extent the disappointments met with in draining heavy land, and explains almost entirely the failure of any sort of drainage to function efficiently in clay land for really long periods such as are common in open soils. But there is little doubt that periodical mole draining is very beneficial in that it prevents water from lying in and on the surface soil, so that the loss of tilth and condition, the lateness and the coldness which water-logging intensifies, are greatly mitigated. Further, the land dries and warms, and comes into use at an earlier date than it otherwise would.

Revival of Older Drain Systems. From time to time one hears of cases where mole draining has given a new lease of life to pre-existing tile-drain systems lying at depths below those usually associated with mole drains. It is a fair inference from such cases that the tile drains concerned were non-functioning, not because of obstructions within them but because of changes in the clay above them and about them. This in the course of time has settled down again to its original impermeable condition, and the drainage water has found it increasingly difficult to get to the tile lines. The disturbance of the ground incidental to mole draining has restored the connection, and drainage will proceed favourably for a further period. This characteristic of the practice can be effectively exploited by providing tiled mains, sited deeper than the mole channels, as the finishing touch to the operation. Covered with a layer of brushwood and clinker, they function for a long while and will last long enough to endure several repetitions of the mole drawing above them.

If one then accepts this method of draining as only a temporary improvement, it is not without point to speculate as to whether, working on these lines, it would be profitable to employ, even in some of the less heavy and cohesive soils, a skeleton lay-out of tiled mains and maintain above them a mole-drain system renewed at more frequent intervals than is customary. Following up these considerations it is perhaps relevant to touch upon the subject of deep cultivations. What is the effect of these upon drainage, especially in clay land? The impermeability of such land has already been stressed, together with the fact that the free movement of water in it is confined to the tilth layer. Where the land is in grass this layer, albeit a shallow one, is maintained by the combined influence of the grass roots and natural weathering. In arable cultivation, tillages are relied on to keep this layer open. The effect of deep cultivations is to extend this layer downwards and to give percolating water a bigger influence both for good and for evil. If there is a good drainage system below, the water will get into it more easily, and in the long run there will be a greater depth of aerated soil. But with inadequate drainage, or, in the absence of drains, there is inevitably a greater depth of water-logged soil, which takes even longer to dry out than a shallow tilth, and further, is much less firm underneath, which becomes disagreeably apparent when it is desired to work the land. There is evidence that the best balance between deep cultivations and drainage is a delicate one, and it is possible to carry the argument in favour of deep tillages too far in the case of clay land.

Some Causes of Failure of Mole Drainage Schemes

Among the many schemes of mole drainage which were carried out in 1940, it is not surprising that some failures came to light. Much of the work was carried out as a matter of urgency and in acknowledged unfavourable circumstances, and risks were taken which might in happier times have been avoided. The long frost-bound spell of January–March 1940 created considerable arrears of all farm work. It was followed by a period of marked wetness and water-logging of the soil, later in many areas by a marked spell of drought, which was suddenly broken by an excessively wet period in late October and early November, throwing a big strain on all the new drainage systems. The mole plough was to be seen in operation in some quarters throughout the summer, often when the ground was very hard. It may be useful to draw attention to some of the causes of failure in mole drains which have been encountered.

1. *Failure to clean out the necessary ditch thoroughly.* This leads to lack of clearance between outfall pipe and ditch water level, lack of depth of the outfall pipe, inadequate fall in the leads, and so failure to get the water away quickly enough, submergence of outfalls, filling up of leads, mains and moles with water, and finally, rotting, silting and collapse of mole channels.

2. *Drawing moles direct from the ditch* and doing without mains. The loamy nature of the soil at the foot of a slope and alongside of ditches, coupled with the lack or even total absence of fall in many cases, makes this practice highly unsatisfactory. The channels collapse and go out of action very rapidly.

3. *Excessively wide intervals.* The success of mole draining depends on the opening effect which it has on the soil and the ease with which the water can get to the channel. In farm practice 3 yd.

intervals is about the optimum: 10 or 12 yd. intervals are inadequate and mean that water standing or accumulating more than 2 yd. from a mole channel is unaffected by it.

4. *Excessive length of mole channels* or *inadequate allowance of mains.* As the length of a mole channel increases, its ability to cope with the load of water decreases and the chances of deterioration by decay or silting increase. 200 yd. is a good limit to have in mind. In some circumstances 250 yd. is safe, but in general, the less the fall, the shorter should be the mole drains. Land with falls as little as 1 in 300 or even less can be successfully mole drained if tiled mains can be suitably sited so that the mole length served is not more than 100 yd.

5. *Inadequate moled mains.* A big reduction in costs per acre can be achieved by substituting moled mains for tiled mains, but where this practice is resorted to it should be used with care and judgment. The moled main is rarely any bigger than the ordinary mole channel, so that several of them should be drawn to replace a single-tiled main. Three should be the minimum; four or five are better. They should be 3–4 yd. apart, deep enough just to clear the minors, and should be drawn first.

6. *Too few leads to the ditch.* It is a desirable aim to have the drainage of 5 acres as a unit, with the equivalent of a 3 (or better a 4) in. tile as main and outfall. If a given field is served by a long single main, it should have leads to the ditch at the rate of one per 5 acres or 5 chains run of main.

7. *Moling on too steep slopes.* This may lead to rapid erosion in the channels and silting lower down, or the rapid run-off may choke the mains, cause a head of water to appear and increase in the minors until its pressure bursts them just above their junction with the main.

8. *Mole plough out of adjustment.* Bad setting or shifting due to wear may result in the digging effect of the mole plough point becoming excessive. If this happens, the channel is elongated vertically and the bursting effect of the plough becomes harmful to the stability of the mole channel.

9. *Absence of mains when moles are drawn.* Sometimes mole channels begin to function as soon as drawn. If rain comes, the rush of water is tremendous, and the results to the mole channels, if there is no outlet, are deplorable. Accordingly, whatever the final intention, all mole draining should be preceded by the provision of mains, either the actual mains desired, or temporary moled mains.

10. *Too wet or too dry subsoil when moles were drawn* (fortunately not frequent causes of failure). If moles are drawn in early spring, when the subsoil is thoroughly wet and subsequent events prevent that subsoil from drying out before the next drainage season, it may be found that the mole drains are very slow in action. This does not necessarily mean they have broken down, and such a condition may right itself the next year.

In some cases of excessively wet subsoil conditions, the bob or trailer plug of the mole plough may temporarily seal the channel against the slit so that heavy rain soon afterwards may result in collapse of the roof and silting of the channel.

Where the mole plough was forced through a hard dryish soil in late summer or autumn, the chances of collapse or silting of mole channels were greatly increased last autumn, because of the excessive rains of late October and November, causing large initial run-offs in many systems.

11. *Unsuitable soils.* The less clayey the subsoil, or the more sand and silt there is in it, the less stable will be a mole channel drawn in it, and the more susceptible a mole-drain system will be to any of the possibilities outlined above.

MOLE DRAINING MACHINERY

Prior to the outbreak of war in 1939, the steam tackle reigned supreme as a means of carrying out mole draining. The big heavy mole, attached to a long beam supported by a four-wheeled carriage, was drawn by means of a long wire cable working between two steam engines standing at either end of the field. This method of mole draining was formerly a routine operation on clay land in the eastern counties. Working at depths of 24–30 in., and 3–5 yd. intervals, these machines can do their steady 10 acres per day. Though latterly much less in evidence, they still, in 1939, occupied an important place in the farming of that part of the country. A survey of some forty-eight farms in East Anglia covering a total area of 11,000 acres shewed that between 1932 and 1939, 1,400 acres of them were mole drained. Of this total, all but 30 acres was done by contractors, 66 per cent of it by steam tackle, 27 per cent by track-laying tractors with direct haulage, and 7 per cent by ordinary wheeled tractors. The depressed state of farming in the clay areas had caused many sets of steam tackle to be laid up; but according

a

b

Plate XI *a* and *b*. A modern mole plough hauled direct by a heavy
track-laying tractor. (*See p.* 111.)

a

b

Plate XII *a* and *b*. Combining the old with the new; a powerful and effective alliance —a Fowler Steam Mole Plough, from which the front wheels and pulley have been removed, coupled direct to a Caterpillar D 7 tractor on wide tracks (75 H.P.). (By permission of *The Farmer and Stockbreeder* and Mr P. J. Macfarlan.) (*See p.* 112.)

to the Steam Cultivation Development Association there were in England at the outbreak of war 109 steam-plough sets available for immediate use and 49 which could easily be reconditioned. A few of these double-engined sets had been converted to Diesel-engined machines, which freed the farmer of the duty of carrying coal and water.

With the prosecution of the Government scheme for extensive mole draining, the use of the heavy track-laying tractor to haul the mole plough direct has rapidly come to the fore. In the early days of direct haulage by tractors, when they were mostly of the wheeled type and of comparatively low power, lighter forms of mole plough were evolved, which produced a smaller channel, while working at less depth, e.g. 2 in. in diameter at 14 in. depth, which is about as much as a wheeled tractor of 12 drawbar H.P. is capable of doing in favourable conditions. It was hoped by this means to introduce the mole plough as part of the normal equipment of a heavy-land farm and to make it a farmer's implement. But the clay-land farmers of the eastern counties were unimpressed and, so far as they could afford it, continued to employ the steam tackle until quite recently. With the rapid development of high-powered tractors, however, especially in conjunction with track-laying types, their capabilities in mole draining became more apparent, and at the moment their power easily exceeds the maximum capabilities of the standard types of mole plough, which for the most part still fail to attain the calibre and depth of the work of the old steam tackle. The great advantage of direct haulage lies in the compactness and handiness of the tackle, in the elimination of the need for carrying water and coal, and in the fact that it can be handled by one man.

The various types of work which can be obtained are indicated in the following table, which gives approximate achievements in stiff clay subsoils:

Type of tackle	Size of channel produced in.	Depth of work in.
Standard twin-engined steam (or Diesel-engined equivalent), with cable	$3\frac{1}{2}$	24–30
Heavy track-laying tractors, 45–50 H.P., direct haulage	3	24
Track-laying tractors, 25–30 H.P.	$2\frac{3}{4}$	18
Wheeled tractors, 12–14 H.P.	2	14–15

The limitations of the lighter-wheeled tractors have been over-come very effectively by fitting them with a winch and cable. The mole plough is attached to the cable, and the tractor proceeds ahead, paying out cable as it goes. When it has reached the end, it anchors itself by means of a suitable spade attachment, and winds in the cable on its winch, thus drawing the mole plough along. The process is repeated throughout the length of the desired channel. Such machines do excellent work, of a calibre and depth equal to that of the heaviest direct haulage, but their speed of work is low and they can only give a fraction of the daily output of direct haulage methods or of the steam tackle.

The prosecution of the war-time mole-drainage campaign has, as a result of the intensive and continuous use of all existing mole ploughs, revealed the fact that the design and production of mole ploughs has not kept pace with the development of the power units suitable for this class of work. It remains a contractor's operation and on all counts—depth and calibre of work, the ability to draw moled mains up to 30 in. in depth, and the strength to stand up to sustained effort day after day, often in trying conditions—the need for a modern mole plough of the weight and strength of the implements which have served the clay counties so well with the steam engines in the past has been acutely felt. It would appear that the present critical period in our agricultural history super-vened when the efficient steam drainers had mostly fallen into disuse, and before direct haulage had progressed far enough in the matter of heavy cultivations to have produced a satisfactory suc-cessor to them. That this can be done is demonstrated in Plate XII, which illustrates the improvised combination of a powerful track-laying tractor (Caterpillar D. 7) with an old Fowler type of mole plough, evolved and operated with success by Mr W. Mardell, of Hardwick, on the very stiff Gault and Boulder Clays in the neigh-bourhood of Cambridge.

PROCEDURE IN MOLE DRAINING

Approved Schemes. At the present time practically all mole-draining work on farms is being carried out in accordance with schemes submitted to and approved by County War Agricultural Executive Committees. They or their officers examine and approve the schemes in respect of the suitability of the soil, the lay-out, the provision of mains, and the depth and distance of minors.

Start with a Main. In the field the aim of the operator should be to produce a job which will be as neat as possible and in a condition to function from the very start. The work should commence at the outfall end of the system and be built up gradually away from it, so that if for any reason work is interrupted and drainage ensues before it is completed, the water will be able to get away without harming what has been done. There should be a main and outfalls, before the minors are drawn. They need not necessarily be in their final form, but should provide the means for the drainage to get away freely from the beginning. During the past two years it has been possible to see a number of instances in which moles have been drawn and left without mains or outfalls, with the result that they were unable to function when the rain came and so suffered serious harm.

Types of Mains. In general, there are four methods of mole draining. The minors may be drawn over pre-existing tiled mains, over new tiled mains put in specially for the purpose, over moled mains, or direct from the ditch without any mains at all. Where the amount of fall is adequate, i.e. steeper than 1 in 200, it is usual to site the mains alongside the ditch at the foot of the greatest slope or across the slope if the length of the minors warrants a second main; where the fall is less than this amount or the ground has an uneven surface, they are sited in the most convenient low places in which they can be given a fall to the ditch.

Siting the Main. Where the main runs alongside the ditch, its actual position should be just far enough into the field to get it in the lowest point of the minors (which is often a few yards from the ditch) and to give the mole plough room to attain its maximum depth when it crosses the main during the actual drawing of the minors. Whatever the final intention may be, there should be no hesitation in drawing moled mains, before proceeding with the minors, in all circumstances except where a new tiled main is already in position. If reliance is being placed on a pre-existing tiled main, an extra moled main some 3 yd. on its lower side can do no harm and may prove to be a useful auxiliary provided that it is duly linked up with the existing leads. In drawing it, care must be taken not to break any existing pipes. If a new tiled main is intended but has not actually been laid, it is advisable to draw at least two moled mains, the inside one being on the line where the tiled main will eventually come, so that the mole slit may facilitate the digging of the trench. If the scheme is to depend on moled mains alone, these should number three at least, and preferably four or five, some 3 yd. distant from each other.

Depth of Main. The depth of moled mains should be decided in accordance with that desired for the minors; in practice it will generally depend on the power of the tackle employed. It should be deeper than that of the minors by an amount equal to the diameter of the trailing bob, so that the minor channel just clears the roof of the main. Thus with a $3\frac{1}{2}$ in. expander, the mains might be drawn 24 in. deep and the accompanying minors at $20\frac{1}{2}$ in. Obviously, when a tiled main exists, the minors must clear it comfortably, if the mole plough is not to break the pipes where it crosses them. The depth should be ascertained by digging or probing with an auger or rod.

Drawing the Minors. All the moled mains having been drawn, the plough should be reset for drawing the minors, which should be commenced from the outfall end. Thus, as soon as the mole plough has got clear, it will be possible for a lead to be dug to connect the mains with the ditch. If the work is carried out in this order, as much of the drainage system as is completed at any moment can function as soon as is necessary without harm to minors, mains or leads, even if the work is stopped for any reason. Leads should be provided at the rate of one per 5 acres of land served or per 5 chains of main.

Low Places. In drawing mole channels, it should always be in the operator's mind that the plough works at a set depth, and that the channel reflects the irregularities of the surface except in so far as the fitting of a long beam to the plough enables it to produce a more even result. Hollows, old plough furrows and surface drains are difficult to cope with but, as far as possible, the channels should not cross them, or a low place will occur at each such point, causing water to be held up, to the detriment of the mole drain and its stability.

Direction of Pull. There is a definite though small advantage in drawing all channels uphill; against this practice is to be set the time lost in running back empty and the greater expense of such work. At the moment, time is a vital factor in drainage work, but this is no justification for tearing through the ground at high speeds, regardless of the quality of the work produced.

Drawing from the Ditch. Sometimes moles are drawn direct from the ditch. This is bad practice unless the object is to get the plough to its set depth earlier in its run, a desirable aim where the ridges come close to the ditch so that the width of even ground or headland in which to draw the mains is rather restricted. Where moles are drawn from the ditch and no main at all is provided, the channels quickly collapse and go out of action, even if care is taken to start the plough at the full depth. The weakness of the soil, the almost invariable presence of a spoil bank and the irregularities of the ditch sides all conspire to make such mole channels unsatisfactory and short-lived.

Ridge and Furrow Land. The existence of ridge and furrow, especially where it takes the form of high-backed lands, must be allowed for. These ridges vary in width from 4 to 18 yd. and in height from 5 to as much as 30 in. If they are narrow and shallow, one channel along each furrow will suffice, but where the width is greater than 7 yd. there is need for more than one. Wider ridges are usually deeper, so much so that moles along the crest do not function, the water running down the steep sides into the furrow. At the same time the area of the ridges is too much for a single channel to cope with so it becomes advisable to draw three moles along each furrow, one in the centre and one on either side of it some 2 or 3 yd. away, according to the width of the ridge.

When to Mole Drain. Land is in the best condition for mole draining when the surface is dry and firm enough to take the tackle without much harm to its tilth or structure and at the same time the subsoil is moist and plastic enough to give a smooth channel behind the mole plough. If it is too dry, the effort of drawing the mole plough becomes enormous, the wear of and damage to the tackle is excessive, while the bursting and fissuring effect of the mole in the neighbourhood of the channel is so pronounced that the channel is less stable and will have a shorter effective life. On the other hand, if the soil and subsoil are too moist, the greater ease of working is offset by the fact that the channel may be sealed off from the slit by the pressure of the expander, the fissuring effect above the channel may be less pronounced and more easily obliterated by subsequent settling, and the resulting work less effective. If subsequently the drainage water has greater difficulty in entering the channel, the clay in which it lies will be wetter, softer and more prone to collapse, with more rapid decay of the whole system. If

8-2

water is actually standing on the surface or in the topsoil, it will at once rush into the slits and fissures after the passage of the plough, carrying with it much mud and silt which is bound to have a bad effect on the newly formed channel and the fissuring in the ground above it. The best results can be expected when work is done in ground free from water in or on the soil and is followed by a period which is moderately dry or at all events free from rain, so that the channels have time to set and consolidate before they are called upon to carry drainage water.

It will be seen that the right conditions are most likely to occur in the late autumn, spring, or early summer. In this country, as a rule, the summer months cause clay soils to dry out gradually from the surface downwards, sometimes perceptibly to a depth of several feet. At moling depth, in such conditions, the clay is too firm or hard for successful moling. When drying ceases with the advance of autumn, the rain begins to re-moisten the soil from the surface downwards, and it may be some time before the clay at moling depth becomes soft enough for the operation. Soon after this happens, the chances are that any heavy downpour will cause the surface to become too wet and the topsoil water-logged. This condition will continue intermittently, according to the weather, until the spring.

In March or April it can be expected that the drying of the ground will soon give a firm surface, free from water-logging, with a moist plastic subsoil condition eminently suited to mole draining. This may continue until May or even June, with the rest of the summer and early autumn for the channels to consolidate. Accordingly, it can be said that the best time for mole draining is most likely to occur in spring and early summer, or in late autumn, but with the possibility of the right conditions being found occasionally during winter months also, depending on the weather.

CHAPTER XI

THE WAY IN WHICH DRAINS WORK. RATES OF FLOW FROM OUTFALLS[1]

From the nature of the problems involved, field drainage does not lend itself readily to direct and precise experiments. In the case of tile drains in open soils some of the difficulties to be overcome before any comparison of the effect of different methods can be made are obvious. Efforts have been made to tackle some of the problems of mole draining by direct experiment, and it has been demonstrated that there is little to fear in the way of one plot affecting events on the neighbouring one. This fact, in conjunction with the small intervals at which mole drains are usually placed, makes plot comparisons of the effect of varying depth, distance, and age, possible. Apart from the possibility of direct experiment, much is to be learned, however, by the observation of actual working drains in the field. Unfortunately, records of such observations are not plentiful. About the middle of last century numerous papers on field-draining practice and problems appeared in the *Journal of the Royal Agricultural Society of England*. Among them is one by J. Bailey Denton, containing the tabulated records of rainfall, outfall, level of the water table, barometric pressure, soil temperature, and air temperature, over a period from October 1856 to May 1857, on a new field-drainage system laid down by him on the Hinxworth Estate in Hertfordshire. His outfalls covered a range of soil types, of which the lightest and heaviest are discussed here. The character of the soils and the method of draining them can be summarised as follows:

Soil parent material	Gravel drift on lower chalk	Gault clay
Analysis by J. T. Way	Sands and clays, 24·4 % Carbonate of lime, 68·3%	Clay, 63·3 % Carbonate of lime 32·4%
Condition of land	Lay very wet before draining	Very stiff and impenetrable
Type of drains	Wide parallel, 4 ft. 4 in. to 4 ft. 11 in. deep, 57 yd. to 59 yd. apart	Close parallel, 4 ft. deep, 25 ft. apart

[1] Incorporating by permission of the Editors extracts from the *Journal of Agricultural Science*, Vol. XXIV, 1934, pp. 349–367.

The records constitute a detailed account of events during the whole drainage season, of which the main features are shewn in Fig. 25. Bailey Denton reviewed his figures on a monthly basis and drew attention to the gradually increasing clearance of excess water as the winter advanced. On this view he found that more water was cleared from the light land than had fallen on it as rain in the months December and February, and that this was the case on all the recorded fields in February. The writer has examined the figures on a different basis. When the rainfall and outfall records for the year are plotted, it is seen that alternations of wet and rainless periods are accompanied by marked flow and ebb in the rates of outfall. Table V presents the data in periods of which each begins with the onset of one flow and ends with the onset of the next, i.e. during the more active part of the drainage season; the

TABLE V. *Wet periods and resultant outfall as shewn by J. Bailey Denton's records*

	Rainfall		Light-land outfall		Heavy-land outfall	
Period	In.	Gal. per acre	Gal. per acre	As percentage of rainfall	Gal. per acre	As percentage of rainfall
Oct. 6th–31st	1·34	29,660	12,800	43·2	0	0
Nov. 1st–25th	1·07	23,970	7,630	31·8	0	0
Nov. 26th–Dec. 4th	0·56	12,540	7,600	60·6	520	4·1
Dec. 5th–29th	1·62	36,290	25,590	70·5	5,230	20·4
Dec. 30th–Jan. 8th	0·54	12,100	7,900	65·3	3,950	32·6
Jan. 9th–20th	1·18	26,520	19,070	71·8	15,520	58·6
Jan. 21st–Feb. 5th	0·77	17,230	23,840	138·3	14,120	81·9
Feb. 6th–28th	0·18	3,970	22,120	557·9	7,820	197·2
Mar. 1st–31st	0·82	18,550	*8,420	45·5	3,310	17·8
April 1st–30th	1·44	32,570	6,700	20·6	6,180	19·1
May 1st–31st	0·75	16,970	4,180	24·7	3,420	20·0

* At this date a new field was substituted in the records.

months March, April and May, unaccompanied by marked variations in outfall, are left as such. The results are presented graphically in Fig. 26 in conjunction with the movements of the water table on the two fields, as assessed by daily measurements in test holes dug between the drains.

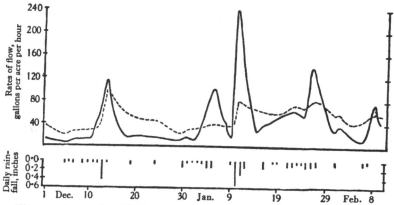

Fig. 25. A sample of Bailey Denton's records of outfall performance of tile-drain systems, based on regular daily measurements.

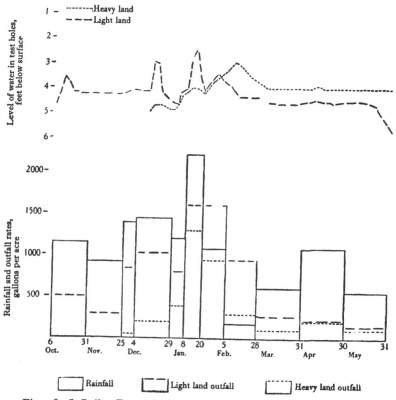

Fig. 26. J. Bailey Denton's drainage records—Hinxworth, 1856–57.

From Table V it can be seen that the soil became gradually less able, as the winter advanced, to absorb or hold all the rain which fell on it, and a steadily increasing proportion of each fall apparently ran off via the drains. The figures for the heavy land in this respect are much more regular than those for the light land. As one flow followed another the drains became less able to clear the excess before the next flush occurred, until during the periods January 20th–February 5th, and February 5th–28th, the light-land drains ran off more water than fell on them as rain, the heavy-land drains shewing a similar but less marked tendency. By the beginning of March the position had substantially improved, a dry February having enabled the drains to deal with the accumulated excess and the soil to absorb more of the rain which fell on it. It must be remembered that the light land lay wet before draining and that it very probably received water from neighbouring higher ground, so that the rainfall cannot be safely regarded as solely responsible for all the water with which the drains were called upon to deal.

An examination of the data respecting run-off in conjunction with movements of the water table reveals some interesting points. It must be noted that the light-land drains ran continuously from October 1st, while the heavy-land drains did not commence until November 27th. From Fig. 26 it is apparent that the onset of each flush of the light-land outfall is associated with a marked though temporary rise in the level of the water table and a subsequent fall to about drain level. Occasionally this was followed by a period of stability until the next wet spell. This was not the case, however, during January, when several wet spells followed one another quickly, causing further upward movements of the water table before it had time to subside to normal, and involving the presence of a surplus of water in the soil to be cleared later when the next rainless period occurred.

The test holes on the heavy land told a different story. After December 9th, each wet spell caused a marked rise of the water level in the test holes, but no marked fall took place until February and even so there was no sign of stability until March 2nd, from which date the level was approximately the same as that of the drains. The fluctuations in outfall rates, however, were frequent and considerable, and one is driven to the conclusion that there is no connection between the apparent level of the water table and the performance of the drains. These observations on the heavy land can be explained by surface run-off through the comparatively permeable topsoil and the disturbed earth filling the drain trenches.

These records were the subject-matter of protracted discussion before the Institute of Civil Engineers. This body of opinion found it difficult to credit the run-off of about three-quarters of the rainfall during the winter as shewn by the light-land outfall records, and no data were available to indicate to what extent the figures were influenced by inflow from neighbouring higher ground.

Bailey Denton made his outfall measurements with a stop-watch and buckets graduated in pints. His critics considered this apparatus inadequate, but made no comment on the limitations of single daily readings. Daily readings have proved adequate to secure a good general idea of events in the working of field drains on various types of soil during a whole season, to measure the effects of a succession of wet and rainless periods, and to give an idea of the capabilities of individual drainage systems. The more marked fluctuations connected with heavy land, however, point obviously to the need for closer observations of outfall performances. The need for self-recording drain gauges has been realised for some time, but the difficulties of producing and erecting meters to suit field-drainage systems and to cope satisfactorily with tremendous fluctuations of flow have proved by no means easy of solution. The nature of the whole problem of drainage of necessity involves very little space between the outfall pipes and the ditch into which they empty. Promising apparatus has been devised both by H. Janert and by J. H. Blackaby. Records obtained by the latter shew the type of fluctuations which occur from hour to hour and the way in which the drains can respond to individual showers of rain.

The behaviour of outfalls is influenced by many factors, some of which are by no means obvious, and of which now one, now another, will be the more potent. With general geological, physiographical and climatic conditions we are not at the moment concerned. In the case of a particular parcel of ground it is the texture of the soil, as reflected by the size distribution of the mineral particles together with the amount of organic matter, which decides its general drainage properties. In normal circumstances, weathering results in the aggregation of these constituents into crumbs with the production of a definite soil structure. The structure of a soil is not uniform throughout its profile and is not immutable. Annual seasonal changes take place but even they display variations from one year to another. Different treatments have different effects, so that the recent agricultural history of any field has an important influence on its drainage properties at any given moment. In particular, the permeability and water-absorbing power of the worked soil, and of the subsoil too, are considerably influenced by previous treatment. The total amount of rainfall for the season, the manner in which it is distributed over the season, and the intensity of individual falls of rain, all exert an effect on drainage results. With so many variants in operation, it is a difficult matter to trace the influence of any one.

Since 1930 observations have been in progress on the Cambridge University Farm in connection with mole-draining investigations, with the object of obtaining more accurate information on the performance of drains and the factors which influence them. All the observations were made on heavy gault clay soil, on drainage systems freshly laid in the ordinary course of farm management. Mole draining with steam tackle was the means adopted, with the channels 20–24 in. deep and 9–10 ft. apart, except on grassland, where sets of three were drawn in the bottom of each existing furrow. The mole systems were provided with collecting mains of tiles leading to piped outfalls. These were watched as closely as practicable, particularly after falls of rain, and measurements of the outfall rates were made by means of suitable collecting vessels and a stop-watch. A sample of the records thus obtained is shewn in Fig. 27. The vagaries in the performance of different drain systems, i.e. in different parcels of land, are indicated in the following account. Outfall *Z* was serving 3·9 acres of old grassland, with a general fall of 1 in 80, in high-backed ridges, with three mole drains in each furrow; *B* served 6·5 acres of arable, with a fall of 1 in 120, with mole drains at 3 yd. intervals; *E* served 5·2 acres of similar arable with a fall of 1 in 110; while *M* served approximately 15 acres of arable with a fall of 1 in 300.

The field in which outfall *Z* was situated was mole drained towards the end of September 1931, and the areas serving *B*, *E* and *M* in the early months of the same year. October was a rainless month, but with November wet weather arrived. On November 7th, after 0·4 in. of rain during the preceding day, *Z* began to run at 9 a.m., rose to a rate of 50 gal. per hour and ceased by 2 p.m. No others ran. On November 9th, it began to rain in the morning,

but no drains had commenced to run at 3.30 p.m. *Z* began soon
after 4 p.m. and within 50 minutes was running at 333 gal. per
hour; by 6.15 p.m. this had fallen to 98. On the 10th and 11th
other small flushes occurred. A fall of 0·49 in. on the 18th affected
all drains to a marked degree, though the rates and their changes
varied greatly. The rain ceased for a time at 7 p.m. and *Z* attained
its maximum output 2 hours later. Further rain during the night
brought *B* and *E* into action, their rates slowly increasing to a

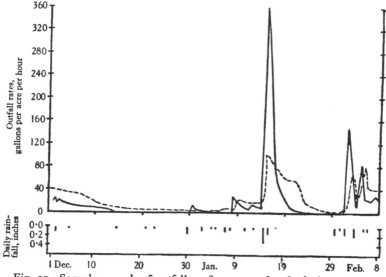

Fig. 27. Sample records of outfall performance of mole-drain systems,
based on measurements at intervals of 15–30 minutes.

modest maximum during the following day. Within 3 days *Z* was
dry again, but the others, on arable land, continued to run gently.
A fall of 0·15 in. on the 26th, followed by 0·3 in. on the 27th,
produced a marked response, *Z* on this occasion increasing to
almost 4,000 gal. per hour by 5 p.m., from 94 at 1 p.m. The arable-
land drains once more came into action later and more slowly,
their rates increasing during the night. January, February and
March 1932 were comparatively rainless months with few single
daily falls exceeding 0·1 in. There was, as a result, very little activity
on the part of the drains. April and May, however, were extra-
ordinarily wet with heavy individual falls of 0·2–0·5 in. within

24 hours, every few days. These were accompanied by frequent flushes from the drains with which it was impossible to keep pace owing to the high rates of flow, the frequency of the fluctuations and repeated inundation of the outfalls. On May 3rd Z was running at 4,100 gal. per hour. This period of excessive rains ended on May 28th and from then until June 30th there was practically no rain. 1·2 in. of rain, spread over the last 3 days of the wet period, kept all the drains running strongly, but within 4 days of its cessation those on the grassland were dry, while those on the arable land continued for a further 7 days. The June drought was ended by the incidence of 0·61 in. in the early hours of July 1st, but Z was the only outfall to shew any measurable response. In about 1 hour of flow it rose from 0 to 70 gal. per hour and receded to 0 once more. On the arable land only one or two outfalls responded, and then only with a trickle of short duration.

Considering the freshly moled grassland area served by Z, two points strike one immediately. The clearance of individual falls of rain took place quickly and the intensity of flow increased with successive falls of rain. The arable land cleared much more slowly and the run-off was much less intense. The outfall records for Z and E are summarised for comparison in Table VI, giving outfall rates in gallons per hour, and are diagrammatically expressed in Fig. 28 as run-off in gallons per acre per hour.

TABLE VI. *Summary of outfall records for grassland, Z, and arable land, E*

Date 1931	Rainfall in.	Fluctuation, Z gal. per hour	Period of run Z (hr.)	Fluctuation, E gal. per hour	Period of run E (hr.)
Nov. 7th	0·40	0–50–0	5	0	0
9th	0·21	0–333–98	6	0	0
18th	0·49	0–1440–8	57	0–75–15	60
27th	0·30	94–3870–50	75	148–1070–192	120

The manner in which the response to individual falls of rain is determined by preceding events is well illustrated by these data. The condition of the soil on November 7th was obviously such that a fall of 0·4 in. was easily capable of absorption by the soil. A very small fraction of it was able to percolate into the drains in the grassland, probably by quick penetration through the fissured ground about the newly drawn mole channels, where percolation could proceed more quickly than absorption by the soil. Two days later the balance in favour of percolation was more marked and the smaller fall of 0·21 in. gave a bigger run-off. Nine days later the next heavy fall of 0·5 in. gave a very much larger amount of

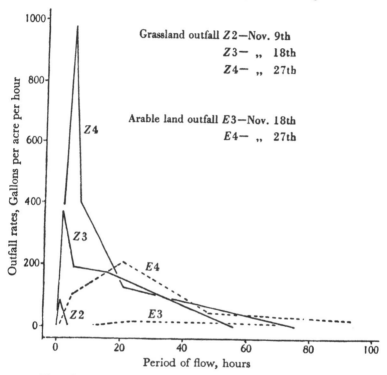

Fig. 28. Comparison of run-off from grass and arable land, on successive occasions.

percolation and run-off. By this date, however, the soil was becoming fairly moist and percolation was probably taking place from farther afield, so that the flow continued for a longer time, and after a smaller fall of 0·3 in. on November 27th, the resultant outfall was both more intense and longer in duration. The arable land above E, with its recently formed tilth, was able to absorb the falls of rain completely on the first two occasions, and it was only on the third and fourth that any excess became apparent as run-off.

In the above presentation of the data (Table VI) no adjustment has been made for differences in the area served by the outfalls. By plotting the records of various complete flows, determinations of the total run-off were made in a number of cases, so that with a knowledge of the various acreages served by the outfalls, a closer comparison can be made, as is shewn in Table VII.

TABLE VII. *Percentage of rainfall evacuated by outfalls*

Outfall	B	E	Z	W	V
Area served (in acres)	6·5	5·2	3·9	3·3	1·3
Cropping of soil	Winter beans	Winter wheat	Permanent grass	Permanent grass	Permanent grass
Discharge as percentage of rainfall (1931):					
Nov. 15th–22nd	6·9	3·1	48·9	—	—
Nov. 22nd–Dec. 6th	49·2	73·5	83·2	—	—
Jan. 13th–23rd	25·4	31·0	54·3	53·6	49·0
April 30th–May 4th	—	53·0	98·0	85·0	—

It will be seen that the trend of these figures is similar to that of Bailey Denton's records, except in so far as rainless periods supervene, when with mole-drain systems, as a rule, outfall ceases and the soil is able, in spite of low evaporation, to recover its absorptive powers to quite a considerable extent. In the instances quoted, the periods December 6th–28th and January 16th–March 22nd were practically rainless. April was a wet month throughout, so that the soil once more became saturated and the rainfall was again largely cleared via the drains, more so on the grassland than on the arable. Outfalls *Z*, *W* and *V* serve different portions of the same field and, in spite of variations in the general slope of the ground, the results show a fair measure of agreement.

In examining the outfall records over the whole season, and comparing events with those dealt with in Bailey Denton's records, certain general points of interest emerge in respect of the various types of drain dealt with. All are characterised by a number of definite flushes, i.e. marked rise and fall caused by separate falls of rain. These are set out in Table VIII.

This record is a reflection of the distribution of the rainfall during the drainage season. The flushes are much more marked on heavy land than on light, in mole drains than in tile drains and on grassland than on arable. In the case of the tile-drain records the flow

Season	Soil conditions	Type of drains		Sept.	Oct.	Nov.	Dec.	Jan.	Feb.	Mar.	Apr.	May	June
1856–57	Light land	Wide parallel, tiled	No. of flushes	—	1	2	2	2	0	0	0	0	—
			,, days without flow	—	0	0	0	0	0	0	0	0	—
1856–57	Heavy land	Close parallel, tiled	,, flushes	—	0	1	4	4	1	0	2	0	—
			,, days without flow	—	31	26	0	0	0	0	0	0	—
1931–32	Heavy grassland	Mole-drained furrows	,, flushes	0	4	3	4	1	2	6	9	0	0
			,, days without flow	30	12	19	0	21	22	0	0	0	24
1931–32	Heavy arable land	Moled 9 ft. apart	,, flushes	0	2	2	2	0	0	5	9	0	0
			,, days without flow	30	31	10	0	23	26	0	0	1	20
1932–33	Heavy grassland	Mole-drained furrows	,, flushes	0	2	4	5	5	5	0	29	24	30
			,, days without flow	30	23	18	2	0	0	0	0	1	0
1932–33	Heavy arable land	Moled 9 ft. apart	,, flushes	0	3	3	4	3	2	0	0	20	30
			,, days without flow	30	10	0	0	0	0	0	0	0	0

TABLE IX. *Maximum rates of outfall attained by drains*

Soil conditions	Type of drain	Distance	Depth	Season	Rainfall during drainage season in.	Maximum outfall rates—gallons per acre per hour
Light land (Bailey Denton)	Tiles	59 yd.	4 ft. 6 in.	1856–57	10·05	Oct. 12th–66. Dec. 13th–101. Jan. 27th–82. Feb. 7th–65
Heavy land (Bailey Denton)	Tiles	25 ft.	4 ft.	1856–57	10·05	Dec. 13th–41. Jan. 10th–214. Jan. 27th–115. Feb. 7th–52
Heavy land, arable (Cambridge)	Moles	9–10 ft.	2 ft.	1931–32	12·12	Nov. 19th–33. Nov. 28th–206. May 3rd–355
Heavy land, arable (Cambridge)	Moles	9–10 ft.	2 ft.	1932–33	13·43	Oct. 28th–57. Jan. 17th–95. Mar. 18th–93
Heavy grassland (Cambridge)	Moles	3 per furrow	2 ft.	1931–32	12·12	Nov. 19th–360. Nov. 27th–390. May 3rd–1,025
Heavy grassland (Cambridge)	Moles	3 per furrow	2 ft.	1932–33	13·43	Nov. 22nd–57. Jan. 16th–370. May 7th–800

between flushes was continuous, although generally at a low rate. In the case of mole drains, it was not an uncommon occurrence for the drains to cease running when any period free from heavy falls of rain supervened. In all probability, the existence or non-existence of a water table is the chief cause of these differences, as Bailey Denton's light-land records shew, but it is suggested that, on his heavier types, it may be the case only so far as the disturbed earth filling the drain trenches is concerned. The existence of such conditions to the extent of 1 sq. ft. in every 25 may have been sufficient to keep the outfalls running slowly for a considerable time after general percolation through the soil has ceased.

The speeds of run-off attained by the various types of drain, as shewn by the maximum outfall rates recorded, are instructive. Some of these are set out in Table IX. They are of some importance in connection with the provision of mains of adequate size.

The differences in behaviour between light and heavy soils generally, and between arable and grass land on heavy soil, have been indicated. The observations which are the subject of this account seem also to reflect variations due apparently to differences in the treatment of individual fields in the normal course of farm management, and to differences in the climatic conditions from one season to another.

Thus it has been noticed not infrequently that autumn cultivations have had the effect of helping the topsoil to dry out with the result that the incidence of drainage has been thereby delayed. Where a fallow has occupied the summer months, it very frequently happens that while the whole layer of topsoil in its huge clods dries out absolutely, the moisture in the subsoil is conserved to a remarkable degree, with the effect that in the following autumn, once rain falls in any quantity, the topsoil is wetted up to capacity, the subsoil is unable to absorb any great amount and drainage takes place much earlier than on similar land alongside which has carried a crop and so has dried out to a greater depth. Such a case is well illustrated in Table X. Outfall *A* served an area which was summer fallowed in 1932, while that behind outfall *B* carried a crop of beans. In August it was noticed that while the *B* stubble was dry and much cracked, the fallow area *A* was covered with a rough tilth. The undisturbed subsoil below the fallow was much moister than that below the bean stubble and cracks were not apparent.

TABLE X. *Run-off from outfalls A and B, expressed as percentage of rainfall for the corresponding period*

Period	Percentage run-off from A	Percentage run-off from B
Oct. 11th–20th	14·8	0·5
Oct. 21st–Nov. 21st	43·5	4·0
Nov. 22nd–Dec. 31st	52·5	5·3
Jan. 6th–30th	68·8	34·6
Feb. 26th–Mar. 16th	65·6	39·5
Mar. 16th–24th	49·7	45·2

Both outfalls began to run on the same day, but while *B* ran only slowly until January 14th, *A*, serving an area roughly only three-quarters of that served by *B*, flowed freely from the start with several periods of comparatively heavy efflux before that date.

Since these observations were made, Dr E. C. Childs, a colleague of the author, has perfected an automatic recording drain meter of a simple type which has been used for measuring outfall rates in conjunction with an automatic recording rain gauge, in some experimental drainage work in the neighbourhood of Cambridge. An example of the records obtained by its means is shewn in Fig. 29. With apparatus of this kind it is possible to examine the behaviour of drain systems with considerable accuracy.

The important factors, then, influencing the rate of run-off of the surplus water from drained land are (1) the moisture content of the soil profile relative to its moisture-absorbing capacity, (2) the permeability of the soil, (3) the amount of fall or slope of the surface, (4) the intensity of the drainage system, i.e. the distance between drains, and (5) the rate of incidence of rain. The effect of a fall of rain on an impervious non-absorbent surface such as a roof or a pavement is to produce an immediate run-off; on an impervious but absorbent soil such as clay, if it is not moistened up to capacity, there may be no run-off at all; on a similar soil moistened up to capacity there will be a rapid response in drainage, completed in a few hours after the cessation of rain—indeed, if the land is water-logged in addition, there will be an immediate surface run-off; while on a permeable soil, percolation proceeds steadily downwards and laterally to the drains, and the run-off is spread out over several days, being maintained by the gradual flow of water from the water table which has risen between the pipe lines.

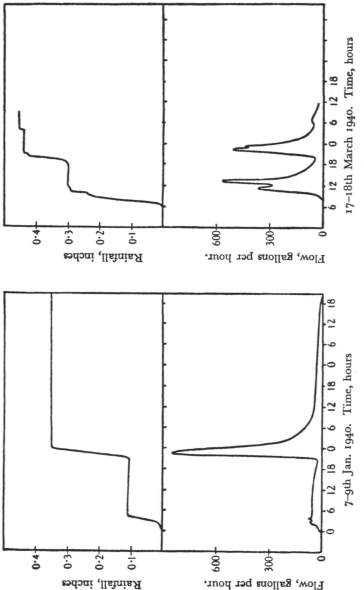

Fig. 29. Records of outfall performance by Childs' self-recording meter, in conjunction with a self-recording rain gauge.

The Provision of Main Drains and Outfalls
(*From a leaflet by E. C. Childs and H. H. Nicholson*)

Mole Draining. The need for providing adequate mains and outfalls for mole drains is well recognised. It is not always easy to decide whether the proposed arrangements in any scheme are sufficient or not. It is nevertheless possible to put forward some general considerations for guidance.

An adequate main drain is one such that water is never seriously held up in the mole channels because of the inability of the main to clear it fast enough. Tables are available shewing the maximum rate of flow in unglazed tiles of any given size, laid with any given fall, so that it is possible to read off the size of pipe needed to cope with a given rate of flow. This cannot be done for moled mains because of the variations produced by soil conditions and tackle and the indefinite nature of the joint between minor and main. In the case of well-laid tiled mains the connection is more perfect.

Information is now available on the subject of run-off from clay land of the type in which mole drains are commonly drawn; it is now possible to assess the maximum rate of flow likely to be forthcoming from any given area and the probable duration of this flow, which follows very closely the rate of rainfall responsible for it. The weather usually experienced during a period of drain activity is a settled drizzle interspersed with heavy rainstorms which may last 1 or 2 hours. These storms contribute the greater part of the drainage water, in flushes which only last 3 or 4 hours, succeeded by a tailing off at quite low rates of flow.

Continuous records from mole drains near Cambridge shew no case where the rate of flow from heavy grassland has exceeded 1,500 gal. per acre per hour. Intermittent measurements over a number of years in another instance have shewn maximum observed rates of 970, 800, 750 and 930 gal. per acre per hour, but none of these was got by automatic recording. On similar heavy land in arable cultivation, the maxima over the same period have been 320, 254, 290, and 231 gal. per acre per hour. It would appear that a main, to be able to cope with all the demands likely to be made on it, would need to be capable of taking 1,500 gal. per acre per hour, in the case of grassland, and about one-third that quantity in the case of arable land.

The following table gives the maximum rate of free flow in gallons per hour from pipes of various sizes, laid with a number of different falls:

Fall ...	1/500	1/300	1/200	1/100	1/50
Diameter of tile					
2 in.	250	300	380	530	750
2½ in.	450	580	710	1,000	1,480
3 in.	800	1,000	1,200	1,700	2,400
4 in.	1,700	2,200	2,700	3,900	5,400
6 in.	5,400	6,900	8,500	12,000	17,000
9 in.	16,500	21,400	26,000	37,000	52,500

It will be seen from this table that a 4 in. main with a fall of 1 in 200 will be able to carry the maximum load to be expected from about 4 acres of ordinary clay arable land and would rarely be unequal to the task of catering for 5 acres. On the other hand, a 3 in. tile, with a maximum free flow of 1,200 gal. per hour, would occasionally be running to capacity when serving only 1 acre of freshly moled grassland. A 3 in. main with a single outfall, then, is by no means generous provision for 5 acres of moled arable land, if the aim is to prevent hold-up of water in the main and in the lower parts of the mole channels themselves. The serious harm which may result is bursting of the mole just short of the main or somewhere in its lower parts. In grassland it may be considered that there is less likelihood of breakdown or of bursting because of the binding effect of the turf.

Consider the simple case of a 15-acre field, rectangular, 15 by 10 chains, with moles parallel to the shorter side and one main along the lower longer side. Here 1 chain of main serves 1 acre of land, an average figure in mole draining. Applying the above considerations, it will be found that a 3 in. tiled main would need outlets at intervals of 2 chains, or seven to eight in all, to prevent the possibility of hold-up, but a 4 in. tiled main would have sufficient carrying capacity if it had one lead per 4–5 acres, i.e. three or four in all.

Here then is a simple standard in the light of which to view proposed layouts of mains in mole draining—one main with one outfall, using 4 in. tiles, per 5 acres, to be modified as thought possible in the light of the fall and of the depth and type of the topsoil.

Tile Draining. In open or permeable soils the nature and speed of percolation is entirely different from that in clay soils, and it is to be expected that different rates of run-off will be encountered. Described in general terms run-off is more even, is much more prolonged and is not characterised by marked flushes. There are not many records of tile-drain performance available, except those by Bailey Denton in 1859. His maxima are given below (from one year's daily readings), both for the outfall and per acre:

Area and soil	Layout of drains	Maximum outfall rate	Size of outfall used, in.
46 acres of mixed soils	Part with occasional drains and part with wide parallel drains 20 yd. by 4 ft. 6 in.	4,600 gal. per hour or 100 gal. per acre per hour	8
18 acres light and heavy soils	As above	2,700 gal. per hour or 150 gal. per acre per hour	5
26 acres medium and heavy soils	Wide parallel drains	6,000 gal. per hour or 250 gal. per acre per hour	5
29 acres heavy land	Close parallel drains 8 yd. by 4 ft.	6,800 gal. per hour or 215 gal. per acre per hour	7

Viewing these records in the light of the table of maximum rates of free flow, it will be seen that the outfall pipes used were equal to all the demands made on them, except perhaps in the third instance, where a 5 in. outfall has no great margin in dealing with 26 acres. Such measurements as have been made at Cambridge indicate maximum rates of flow for tile-drain systems of 400–500 gal. per acre per hour. On this basis a 3 in. tile should cope with 2–3 acres, a 4 in. tile with about 6 acres and a 6 in. tile with about 18 acres, at falls of 1 in 200.

It is not uncommon to hear a farmer, pleased with the result of his drainage work, enthuse over its performance in terms such as 'running full bore' or 'hitting the opposite side of the ditch and punching great holes in it'. In many cases the criteria may be more graphic than accurate; but sometimes this stage of performance is actually attained. It is no cause for complacency, however. A pipe running full bore is overloaded, a condition fraught with danger to the system behind it. The accumulation of water to this extent produces a head of pressure which may become big enough to cause bursts at weak points in the system.

Occasionally an outfall is seen to run jerkily, or to pulsate in a way which suggests the existence of an airlock or a partial syphoning action in the pipe. It was probably observation of this phenomenon which led to the old-established practice of providing mains in particular, and sometimes minors too, with air inlets at their upper ends; but in this instance also the cause must be sought in the inadequacy of the main for the load which it is called upon to carry.

CHAPTER XII

DRAINAGE OF AERODROMES, SPORTS GROUNDS AND FIELD FORTIFICATIONS

Of recent years much energy has been devoted to the draining of land utilised for aerodromes and sports grounds; drainage problems have also become acute in connection with air-raid shelters, dug-outs, and field fortifications. The following extract from a newspaper of May 18th, 1941, illustrates some of the unforeseen results of bad drainage conditions.

WATER IN SHELTERS[1]

Thousands Affected by Court Decision

Many thousands of persons who had raid shelters built in their gardens, and builders who constructed them, are affected by an interesting Court decision.

It is to the effect that, in the absence of any stipulation to the contrary, a shelter should be not exactly waterproof or watertight, but so constructed as to be reasonably free from water through seepage from the ground or otherwise.

There is, the Court held, an implied term in the contract that the shelter should serve the purpose for which it was intended, and that if, through water, those purposes are not served, then the owner may resist any claim for payment.

In the case before the Court the facts showed the shelter was built in dry weather, and, while the weather was dry, the shelter was serviceable, but when it rained, or the soil became waterlogged, water to a considerable depth made the place unusable.

The builder lost his case, and on a counter-claim the owner was allowed a sum for removal of the structure and for restoration of the garden to its original condition, and also a nominal sum for breach of contract.

In the early days of shelter construction very little attention was paid to the possibility of infiltration of water, and local authorities as well as private individuals have found their shelters of no use and have scrapped them.

Other shelters (mostly of the surface kind) have been built by the authorities, and private individuals have put in pumps or taken other steps regarding the places.

[1] *News of the World* Special, 18th May 1941.

In all these various fields of activity the medium concerned and the fundamentals of the subject are the same as those which have concerned the farmer for centuries. The problems themselves differ from those of the farmer chiefly in the efficiency expected of or hoped from the measures carried out and the use to which the ground concerned is to be put. The agriculturalist carries out drainage projects to render his soil more favourable to plant growth and to enable him to stock it or to cultivate it without harm to its fertility. He realises that drainage alone cannot accomplish this and that he must allow time for evaporation to supplement the effects of drainage. He refrains from admitting his stock or putting his implements to work until the ground is dry enough to prevent puddling or poaching. But in the case of landing grounds or sports fields, the demands of the users are so urgent that the time necessary for evaporation cannot be spared, so as soon as the surplus water has disappeared, the ground is put into use, with results which, to say the least, have a decidedly adverse effect on its drainage properties.

It is obvious, too, that quicker get-away of surplus water is desirable, so that there is scope for much more intensive drainage systems than are customary in farming practice, e.g. shallower drains, the use of gravel or hard-core filling above them, and narrower intervals between minors. All these are merely extensions of or intensification of ordinary agricultural practice.

The subject of the drainage of aerodromes has recently figured in the Ninth Report from the Select Committee on National Expenditure, dealing with Air Services. Its remarks are quoted here.

Selection, Acquisition and Construction of Aerodromes

Vast expenditure has been incurred and is still being incurred in the construction of aerodromes throughout the country.

The Air Ministry are responsible for the selection and preparation of sites for aerodromes. The Sub-Committee have taken evidence from contractors, engineers, surveyors, land agents and other persons as well as from officials of the Ministry. They sought this evidence because they have received, and are still receiving, complaints that sites are chosen which are unsuitable or unnecessarily expensive, or that work on these sites is not being executed in a proper manner. In some cases it has been alleged that the land was water-logged and that drainage would be impossible; in other cases that valuable agricultural land had been requisitioned.

Factors Governing the Choice of Aerodrome Sites

In their search for aerodrome sites the Air Ministry are limited by the following factors:

(*a*) strategical requirements;
(*b*) air congestion;
(*c*) the existence of permanent obstructions, such as hills, towns, railways, rivers, main roads and the main electric grid;
(*d*) local meteorological conditions;
(*e*) area of land required;
(*f*) ground contours, soil and drainage.

In elaboration of (*f*) the report says:

The whole of the site for an aerodrome must be comparatively level or capable of being levelled by grading rather than by excavation. An aerodrome should also be well drained and on light soil. This was most essential when grass aerodromes were still largely in use, as many soils are incapable of supporting a sufficiently good grass surface without special preparation and treatment for a long period. Now, when as a general rule concrete runways are being put down for all operational stations and an entire grass surface is only required for training aerodromes, the nature of the soil and drainage difficulties are not deemed to be of so much importance.

In addition to the factors of natural and artificial obstructions, soil, drainage and weather conditions and to considerations of strategic requirements and urgency, the Air Ministry must also have regard to objections from other departments. They are frequently opposed by the Ministry of Agriculture on the ground that they are proposing to take valuable agricultural land.

It is to be noted that soil and drainage are mentioned last of the factors to be considered in choosing a site, but there can be no doubt that these two points exert a telling influence on the operational efficiency of an aerodrome.

Some of the special difficulties and possible sources of trouble in this sphere of drainage work can easily be envisaged. The need for an even surface of considerable extent means that pre-existing ditches must be piped and filled in; ridges and furrows must be obliterated; excrescences must be removed and hollows filled in. Existing land drains may or may not be broken in the process, especially where buildings are erected or runways laid down. Whatever is done by way of such operations, surplus water will continue to move along the tracks which it has followed in the past, and wherever the drains are broken or the underground drift of the water is interrupted, there will trouble arise, unless adequate measures are taken to forestall it.

In the sphere of field fortifications, tactical considerations must obviously come first, but a knowledge of drainage conditions in the soil should be of real value to those whose duty it is to site and lay-out such works, with an eye to the comfort and morale of those who may subsequently have to occupy them. The signs and portents of bad drainage conditions are visible even in midsummer; they have already been described. In particular, the digging of slit trenches, strong points, gun pits and so on provides ample opportunity for spotting the indications of a high water table or of impermeability even when there is no sign of water actually present. Thus forewarned, there is no excuse for neglect to adopt suitable drainage measures where these are seen to be necessary.

When the water table rises, any excavation such as a gun pit or an isolated trench will have water in it up to the level which the water attains in the surrounding land, unless a way out is provided for it. It will generally be possible so to adjust matters that the trench functions as its own ditch, by connecting its lower end with the nearest existing field ditch, with due observance of the usual requirements as to fall. This done, the trench will remain comparatively free from standing water. Gun pits can be dealt with in a similar fashion, so that each one becomes as it were the head of a sap which functions as its drain. In clay land the circumstances differ. Isolated excavations in such soil function as sumps; surface water drains into them from the surrounding soil and they slowly but inevitably fill up to within about 8 in. of the surface as winter advances, unless steps are taken to empty them. Such water can, of course, be got rid of quite effectively by means of a pump and little more will accumulate until the next downpour of rain, when the pumping can be repeated. Or, the excavation can be dealt with as in open soils by suitable connection with a ditch. The most difficult situations are to be encountered in bottom land with a high water table. Here the ditch or stream level is the decisive factor and immunity from flooding can only be attained by the erection of breastworks or the construction of substantial concrete works which can be rendered watertight.

As pointed out earlier in this chapter, one of the most serious difficulties to be surmounted with regard to the drainage of any field sites or works, such as landing grounds, camps, or earthworks, is the immediate and adverse effect of their everyday use in wet weather. Trampling or traffic of any sort in such conditions renders the surface of even the lightest and most permeable soils impervious to water, so that surface pools soon appear and persist. The reason is that such traffic, on a wet soil surface, destroys the natural crumb condition, and the mass of mineral particles and the humus becomes homogeneous and continuous. The larger spaces which existed between the crumbs are obliterated and water can no longer percolate through them. Were it not for the traffic the drainage properties of the surface would suffer little or no harm, and, on drying, the soil would be as open and free-draining as before. Crumbiness in the surface soil is the natural result of the development of the root systems of the plants which grow on it and of the alternations of heat and cold, wet and dry, which we summarise as the weather or the seasons.

Anything which encourages the growth and maintenance of a perennial herbage, e.g. a thick and luxurious turf, will lessen or prevent the occurrence of surface pooling in wet weather. It is impossible to get such root action underground without a vigorous growth of herbage on top. Anything which destroys this herbage prevents crumb formation. Where such conditions have been produced and it becomes imperative to deal with surface pools, various temporary expedients may be resorted to. On clay land or other heavy soils which are impermeable from the surface downwards, shallow water furrows can be ploughed or dug by hand to the nearest ditch or outfall. If, however, the trouble is due to surface puddling of an otherwise permeable soil formation, it is possible to open up the soil mechanically and let the water down to percolate away below. An ordinary garden fork can be used to lever up and thus open the surface of small wet areas. Spiked rollers are used to produce a similar result over larger areas. Local sumps can be dug to a depth and of a size determined by the area to be dealt with, and filled with hard core. The surface water can then be led to the sump by suitable water furrows.

CHAPTER XIII

FIELD DRAINAGE, RIVER FLOW AND FLOODS

As recently as 1920 the Board of Trade commented on the conspicuous lack of information on the subject of river flow in this country. No regular gaugings of the flow in rivers seemed to be carried out in any single instance. Even now very little has been done to correlate events in the fields with those in the rivers. The chief concern of the Catchment Boards, at all events until quite recently, seemed to be flood prevention; they were interested enough in the water after it had arrived, but not before. The subject of the water in, on, and under the earth has been considered and dealt with by diverse authorities in terms of their individual primary interests, e.g. water supply, water rights, navigation, fishing, pollution, and flood control. We have a fair knowledge of the incidence of rain on the country as a whole, thanks to long-continued daily measurements. We know much less about its local variations and the rates of its actual incidence (as obtained by self-recording gauges), but of what happens to it subsequently we know remarkably little. Our ideas have been based mainly on conjecture and rough approximations, at all events until the date mentioned. It is doubtful whether any river authorities have actually measured the contributions of each of their tributaries to the main river in conjunction with the field conditions within each sub-catchment area. Very little has been published on the subject, although it is some time since a Parliamentary Committee on Water Resources recommended that the Catchment Boards should gauge their tributaries as well as their main rivers in order to investigate the nature of the catchment with which they had to deal.

This state of affairs is being remedied. Pioneer efforts are those of Professor S. M. Dixon, on 'The Flow of the River Severn, 1921–36' (see *Journal of the Institution of Civil Engineers*, 1936–37) and of Mr W. N. McClean on Loch Ness and the River Dee. The subject was considered by a committee of the British Association and later by a special Inland Water Survey Committee. In 1934 it was urged that drainage authorities should record their river flows as a matter of routine, and some of them have made a start. As a result of the efforts of the Inland Water Survey Committee and other bodies, gauging stations have been established by a number of drainage authorities on their respective rivers.

The Ministry of Health has published two issues of *The Surface Water Year Book of Great Britain*, for 1935–36 and 1936–37. These give records of river flow from some 60–70 stations, scattered over a number of rivers, e.g. 20 on the Yorkshire Ouse, 6 on the Thames, 3 on the Nene, 1 on the Stour and 1 on the Lee. Presumably this work has continued to expand, at all events until the outbreak of war, and will go on expanding in the future, so that there is a reasonable prospect of a much more accurate knowledge of the connection between incidence of rain and run-off in the rivers, and of all the multifarious events between the two. Percolation, evaporation, field drainage, springs, water supply, sewage, town and road storm water, and floods will all be seen in better perspective, and their relative significance will be more accurately gauged, as records of flow from springs, field drains, ditches, small streams, tributaries and rivers are accumulated, in conjunction with a measurement of the area of the geological outcrops concerned and the permeability of the rocks and soils on them.

Professor Dixon, in his paper on the daily observations at Bewdley, drew attention to the following points:

1. The highest flows in an average year occur between December and March.

2. The highest flows in a wet winter occur between October and February.

3. The highest flows in a year with a wet summer occur indiscriminately throughout the year.

4. The highest flows in the driest year occur in January.

He records one summer flood as revealing some interesting points. On May 31st and June 1st, 1924, widespread heavy rain fell in Shropshire and Worcestershire, beginning in the evening and continuing until the next day, with a total fall of 3–4 in. There was no more for some days afterwards. The river began to rise at 6 p.m. on May 31st; it rose rapidly from midnight until 6 a.m. and reached its maximum at 6 p.m. on June 1st. It remained at this level for 2½ days. He calculated that if the 3½ in. of rain had been run off in 1 day, the flood peak would have been 14 times as high. In fact only 48 per cent was run off, so that 52 per cent was to be accounted for by evaporation, percolation to depth, or absorption in the soil. The power of the soil in this respect has already been described (see p. 20). His table of monthly average run-off for 1921–36 is of great interest, especially when studied in relation to evaporation and to the absorptive power of the soil, as they change through the seasons:

	Oct.	Nov.	Dec.	Jan.	Feb.	Mar.	Apr.	May	June	July	Aug.	Sept.
Rainfall	4·11	3·97	3·57	3·70	2·90	2·12	2·72	2·73	2·54	3·40	3·27	3·11 in.
Run-off	1·51	2·46	2·18	3·11	2·25	1·67	1·29	1·08	0·87	0·71	0·84	0·93 in.
Run-off	37	62	61	84	78	79	47	39	34	21	26	30 %

The increasing proportion of run-off as autumn and winter advance and the decrease with the arrival of spring and early summer fit the figures for evaporation quoted earlier (see p. 15) quite well. This is the underlying cause, but it acts through the moisture properties of the soil, creating a deficit there of varying magnitude which in the conditions of our climate can absorb a substantial proportion of the rain which falls, varying according to the time of year and the distribution of the rainfall in respect of time.

Professor Dixon noted that in certain years the run-off exceeded the rainfall for a given month on several occasions between December and March, just as Bailey Denton's records showed for a small-scale unit like a field. As in the latter case the soil and subsoil were permeable, it was suggested that the explanation perhaps lay in water infiltering from surrounding land, but in the case of a river basin it is much more explicable by the comparative slowness with which water moves underground. The time taken for the passage of water from the surface of the soil to a spring at a distance is appreciable, as is shewn in the example mentioned on p. 42. Professor Dixon comments that 'the existence of this condition of augmented flow renders impossible any study of the relation between rainfall and run-off during winter based on monthly periods. In summer there is a much higher rate of loss and little of the rain appears as discharge. If the rainfall and run-off are plotted for the months of June to August, it is found that there is a definite relation between them.' In the discussion on this paper it was suggested that the dry weather flows were very low and that the explanation lay in the loss by percolation along the direction of the old channel of the river, particularly at low stages of flow. The 'buffering' effect of the soil does not seem to be appreciated by workers in these investigations. Let it be repeated that after even short dry spells in summer, heavy soils in particular are found to be deficient of as much as the equivalent of 3 in. of rain; there need be no mystery in the failure of summer rains to appear in the rivers in appreciable quantities.

The connection between field drainage and flooding in rivers has been a subject of debate for centuries. The arguments follow the same lines, depending on the point of view or the nature of the interest involved. A river and its tributaries constitute the natural drainage system of its catchment area. In general the fall or gradient decreases rapidly from the source to the lower reaches, the volume of water steadily increases, the margin of safety in the banks of the channel decreases and the liability of the adjacent land to be flooded increases correspondingly. Floods are a natural phenomenon in the flat reaches of a stream and in its lower parts. Any improvements effected in the channel upstream will accelerate the run-off and throw a greater strain on the lower reaches unless these are correspondingly improved.

To the high-land farmer the river is the receptacle for the surplus water of his land and what happens lower down is not his fault or concern. His drainage has always gone there. If he is moved to express his views, they may be that his efforts at drainage within his own domain are stultified by neglect lower down. The low-land farmer has different ideas. His forbears, by dint of enormous labour and huge capital expenditure, had reclaimed land from the flood plains and marshes; they had set up complicated drainage systems, improved the rivers and concentrated on flood prevention. Their reward was vast expanses of highly fertile land, but dependent absolutely on the continuance of these expensive drainage measures. In hard times, he looks sourly on the water which comes down to him from the high-land farmers upstream, from the waterproofed urban areas and from the tarred roads (continually increasing in area) and thinks and says that the expense of land-drainage work should fall fairly on all whose drainage water is dealt with and not only on those threatened with flooding.

When the subject of improved field drainage on a wide scale arises, one of the first possibilities envisaged is an enhanced rush of water to the rivers, with an increased risk of flooding lower down. Considering the number of factors involved, the way they interlock, and the independent variation of each with seasonal and climatic differences, the possibility of simple and apparently sound reasoning leading to wrong conclusions is a real one. There can be no doubt that field drainage facilitates the flow of water to the ditches and streams; but whether doing so necessarily means a greater risk of flooding is another question.

Whether or no a drain, ditch, stream or river overflows at any point depends on the simple fact of the amount of water arriving at any moment relative to the amount which can get away. That is to say, it is a matter of the peak load rather than the total amount of water in a given interval of time. It is possible for the latter to be increased while the former is substantially diminished.

When Bailey Denton presented the results of a year's records of outfall performance to the Institute of Civil Engineers in 1861, the discussion on these points occupied four evenings. The gist of his views was that field drainage would liberate from the soil amounts of water on the scale of his records; that it would result in augmenting the flow of rivers in the winter; and that unless these were looked after, the amount would be sufficient to flood extensive areas. He failed to give a clear picture of how drains worked; indeed, on some points, his own records shewed that his ideas were not correct. He firmly held to the existence of a free water table in clay just as in open soils, and opined that clays did not discharge drainage until saturated, when for every volume of water falling on top, an equal volume would be forced out below into the drains.

His audience held many divergent views and was eager to express them. They are interesting and worth noting once more. The Rev. Clutterbuck, having lived near the Thames for 30 years, said that the flood water in it came from the areas occupied by the outcrops of the Oxford, Kimeridge and Gault Clays respectively. Near Abingdon, the floods used to reach their highest level within 72 hours, but the under-drainage of the past 20–30 years had reduced this interval to 36 hours. He did not state whether the peak was as high as formerly. He pointed out that the general effect of field drains might be to deplete the supply of water from surface springs, but not to affect that from deep sources; a conclusion perfectly in accord with the facts that drainage deals with surface water or that part of deep water tables which happens occasionally to come within a few feet of the surface. Mr Samuel Sidney, Mr Hawkesley and Mr Lloyd all stressed various aspects of the view that improved field drainage would inevitably mean easier and quicker access for drainage water to the rivers, especially in the winter, with resultant more frequent floods; but it had not occurred to them that easier and earlier run-off might possibly ease the strain later, i.e. the peak flow might be less high and less dangerous. Mr Hawkesley considered that rivers are improved if their flow is rendered more even; but he held that the more the land is drained, the more irregular the flow of the streams must become. Mr Lloyd believed that the total amount of water going into streams and rivers was not sensibly altered by drainage, but there was more in flood times and less in dry seasons. Mr Field drew attention to some important facts: first, that floods are caused by heavy rainfalls with rapid run-off; secondly, that free or permeable soils behave differently from clays; thirdly, that run-off is not rapid if percolation takes place first, but is rapid if it takes place over the surface; and fourthly, that rivers dependent chiefly on springs are less subject to floods. Mr Homersham supplemented this by calling attention to a little piece of investigational work of his own. He compared events at eighteen bridges over rivers, some of which drained areas of London Clay and some of Chalk country. For equal catchment areas the waterways in the Clay country were 3–10 times as large as those in the Chalk; and while the former were often filled to capacity, the latter were rarely seen in that state. Moreover, the summer flow in the latter case generally exceeded that in the former. His view was that the draining of free soils reduced the incidence of surface run-off and produced a more equable flow in such cases, thus tending to diminish floods. But he did not attribute the same effect to the draining of clay. While admitting that it would lessen surface run-off, he held that the greater intensity or concentration of channels associated with heavy-land drainage gives a more rapid run-off and an increased risk of flooding.

An examination of Bailey Denton's records and those made in recent years at Cambridge brings out clearly the difference in the nature of the run-off from field drains in permeable or open soils and in impermeable or clay soils respectively. The former is gradual and protracted over a period of several days, without a marked peak; the latter is swift and in the nature of a flush, with a marked peak of flow, several times as high as that from open soil in similar circumstances, and of only a few hours' duration. A similar contrast is to be found between the performance of tile drains 30 in. deep and 15 yd. apart in clay land and that of mole drains 20 in. deep and 3 yd. apart in similar land. Actual measurements of the surface run-off from saturated undrained clay land have not yet been recorded, but if one may take the nature of that from a concrete or tarred surface as a criterion, and we know this must be almost instantaneous, then we can logically suggest that the contrast between drained and undrained clay land will be similar, i.e. that the run-off from undrained clay will be the most superficial type and will give the biggest peak flows. In short it is possible that field under-drainage everywhere may have the effect of lessening the peak flow of run-off and of minimising flood risk. When land is undrained and water-logged, whether it is permeable or impermeable matters not; the only run-off is superficial and must be of the flush type. No one disputes the facts in the case of 'storm water' and dealing with it in modern urban development. Few would dispute its significance in local river floods below some of our larger towns. Indeed in some quarters land drainage seems nowadays to carry this rather limited meaning. Drainage appears to be in danger of being confused with and treated on similar lines (even if in a different set of pipes) to sewage.

Such evidence as exists on this thorny problem can be interpreted both ways. A grand opportunity to produce a sound answer has just occurred, but it is doubtful whether anyone has been able to take advantage of it. It does not often happen that there is such widespread activity in field draining as is being pisplayed at present, all duly recorded in the archives of the War Agricultural Executive Committees, in all its detail. There may be river-gauging stations in existence which have recorded the run-off from suitable areas of land for a few years beforehand and whose records for the next few years may provide useful evidence of the effect of drainage work within those areas. It is the Catchment Boards who are in the best position to investigate this problem.

There is room for extensive and prolonged investigation of the movements of water on, in, and under the soil before many of these debatable points can be settled, and the need for correlation of such work from the individual field unit down to the sea is fairly obvious. Experimental work on field drainage is extremely difficult, as conditions vary so much from point to point and the isolation of plots to measure the effects of different drainage treatments involves laborious preparations. Meanwhile there is still much to be learned by the closer observation and study of established drain systems in the field. It is obvious, too, that much is to be gained by the co-ordination of all types of drainage work, field drains, ditches, minor streams, rivers and estuaries. They all deal with the same water, they are mutually interdependent and a just balance should be maintained between them.

CHAPTER XIV

FIELD DRAINAGE. PRESENT POSITION AND PROGRESS[1]

During the years which elapsed between the war of 1914–18 and the present conflict, the spokesmen of agriculture and agricultural investigators alike never ceased to call attention to the steady deterioration in the condition of the field drains of this country. Their growing concern on this account was not surprising in view of the fundamental importance of good drainage as a factor in the fertility of the soil. What has been patent to the farming community for years has now been realised by the country at large, namely, that much of our land is not contributing substantially to the food-production effort because of the woeful state of neglect into which its field drains and ditches have fallen.

About ten years ago, an investigation of 1,000 individual farms in the eastern counties, with a total area of 170,000 acres, shewed that in this part of the country 26 per cent of the heavy land, 13 per cent of the loams, and 3 per cent of the light land were greatly in need of field drainage. So far as the heavy land concerned in this investigation is concerned, it is known, as the result of a later survey of the same farms, that only 12 per cent of the area had been re-drained since that date up to the beginning of the war; and as the method employed was mole draining, whose effective life varies between 6 and 12 years, it is obvious that the arrears must have persisted, if, indeed they have not actually increased.

In the absence of any accurate and detailed soil survey in this country, it is impossible to say just what area of agricultural land falls in each of the above categories; but the clay areas, where field drainage is in the worst plight, cover about 4,000,000 acres. Of this almost half is to be found in the arable counties of eastern England. In examining the position of field drainage it is important to take into consideration the more recent history of individual fields. The general deterioration in field drainage has taken place concurrently with a steady expansion in the area of grassland. It is a truism that successful arable farming can only be carried out on well-drained land, so that it is not surprising to

[1] Incorporating an article entitled 'Field Drainage' from *The Times Trade and Engineering*, Feb. 1941, by permission of the Manager of *The Times*.

find that in only too many cases the fields whose drainage was defective or deteriorating were the first to be laid down or allowed to fall down to grass. This is the position in which much of the grassland which is being or has been ploughed out to-day finds itself; and it is almost invariably true in the case of derelict or semi-derelict heavy land which is undergoing reclamation. Some of the circumstances of the heavy land counties in England are summarised in Table XI, based on the annual Agricultural Statistics, published by the Ministry of Agriculture. In this table attention is drawn in particular to the parallelism between the area of clay land, the prominence of wheat, beans and fallow, the effect of an outstandingly wet year (1937) and the need for field drainage as reflected by present efforts. The figures in the first seven columns are derived from the Agricultural Statistics of the Ministry of Agriculture. Those given for the area of land lying on clay formations have been computed from published geological maps, and as a guide to the distribution of clay land over the country are admittedly imperfect. East of a line joining Goole with Portsmouth they can be considered fairly reliable, but west of that line, they do not include such areas as are covered by Boulder Clay; hence the absence of figures for the counties of the north and the north-west midlands.

Land drainage, of course, is not simply a matter of conditions in the soil itself. Surplus water must be got rid of first into the ditches, then into the main streams and rivers, and finally into the sea. To keep the country as a whole in the best condition needs unremitting care and activity in all these directions. Fortunately, the Government has, for a generation now, been consistent in its efforts to improve our rivers, and it is generally admitted that conditions in them have steadily improved, especially during the past 10 years. Matters are by no means perfect even now, but the rivers are in better shape to carry their loads than they were.

The two chief directions in which a prodigious effort is now required are the re-conditioning of field ditches and the drainage of heavy land. The year 1940 witnessed a new and widespread interest in these matters as a result of the various schemes initiated by the Government to encourage drainage work. It is perhaps too early to gauge the ultimate effect of these several schemes, especially as they have not all been in operation an equal length of time. A late 1941 announcement stated that schemes of mole drainage affecting 133,000 acres, at an estimated cost of £265,000; of tile drainage affecting 68,000 acres, to cost £401,000; and of farm ditching affecting 805,000 acres, to cost £1,033,000 had been approved. It is clear from the facts relative to the heavy land alone that there is need for even more rapid progress than has been made so far. With probably a quarter of the 4,000,000 acres of clay land in acute need of field drainage, it would appear that the effort up to that date had covered some 13 per cent of the need.

An examination of the facts relative to the mole-draining effort brings out some striking points. Of the 133,000 acres approved for grant up to September 1941, almost all lay in the eastern or east midland counties of Essex, Suffolk, Cambridge, Huntingdon, Bedford, Northampton, Hertford and Leicester. Smaller areas have been approved in the counties of Buckingham, Norfolk, Kesteven, Nottingham, Oxford, Sussex and Worcester. There are, however, some counties with substantial areas of clay land, where little appears to have been done. In most of them the proportion of grassland is much higher than in those counties where draining work is being pushed ahead. The lack of field drainage may not be so serious in grassland as in arable, but no one would deny the benefits which would accrue in increasing its stock-carrying capacity and its general productivity.

The question comes to one's mind, whether or no this operation of mole draining is widely applicable in the heavy grassland counties. The fact that some farmers in each are taking advantage of the scheme, considered in conjunction with the many successful demonstrations of the method by the Ministry of Agriculture during the years following 1925, would indicate that it is, and it is to be hoped that in the near future the mole plough will be much more in evidence in these areas.

Mole draining is still in the main a contractor's operation, for the reason that the most desirable work is of a depth and calibre which requires fairly heavy tackle. There are, however, lighter forms of mole plough, and such machines might be much more in evidence as farmer's implements to-day than they actually are. The effects of the mole-draining operation are of such a nature that there is a good deal to be gained by making it a routine cultivation; and the actual drainage effect need be very little inferior to that of heavy work, at all events for a short period, if it is carried out over an adequate system of tiled mains. Too little

TABLE XI. *Shewing the distribution of clay land in the English counties, and its effect on cropping and mole draining*

County	Area under crops and grass 4.vi.1937 (acres)	Arable land as percentage of total under crops and grass 4.vi.1937	Arable land as percentage of total under crops and grass 4.vi.1918	Wheat as percentage of arable area 4.vi.1937	Beans as percentage of arable area 4.vi.1937	Bare fallow as percentage of arable area 4.vi.1937	Increase in bare fallow, 1937 as percentage of that in 1936	Approximate area of land on clay formations Acres	Approximate area of land on clay formations As percentage of total area	Area approved for mole draining under Government scheme to 30 June 1952 (acres)
Bedford	237,849	48	46	27	3·3	16·1	74	170,000	56·5	23,330
Berkshire	312,831	38	61	28	1·1	14·7	2	86,000	18·7	834
Buckingham	360,028	19	54	26	2·3	16·6	59	197,000	41·4	11,651
Cambridge	256,620	75	81	27	1·8	10·5	87	120,000	38·3	26,159
Isle of Ely	209,980	81	80	32	0·9	2·7	148	25,000	10·5	171
Cheshire	485,425	32	43	18	0·06	0·5	113	—	—	1,072
Cornwall	608,255	54	60	3	0·03	0·7	4	—	—	22
Cumberland	511,945	33	41	0·5	0	0·1	27	—	—	690
Derby	436,780	16	23	27	0·17	5·2	76	—	—	1,854
Devon	1,125,610	37	46	7	0·13	1·1	19	—	—	141
Dorset	418,708	24	38	18	0·3	5·7	15	80,000	12·9	1,944
Durham	381,608	34	39	15	0·3	3·9	46	—	—	994
Essex	699,722	56	67	27	2·8	10·8	59	602,000	62·0	153,471
Gloucester	610,899	26	37	22	1·2	6·2	42	130,000	16·3	2,744
Hampshire	524,676	45	63	22	0·3	15·0	11	130,000	13·6	622
Isle of Wight	59,313	30	47	19	0·8	6·9	10	6,000	6·4	373
Hereford	435,813	27	34	18	1·2	1·4	16	—	—	58
Hertford	283,167	51	65	31	1·4	12·6	44	111,000	27·6	31,122

Huntingdon	197,500	58	63	33	3·8	14·1	104	137,000	58·8	37,420
Kent	644,841	39	47	13	1·1	5·8	52	175,000	18·0	7,003
Lancaster	686,475	27	39	15	0·05	0·8	8	—	—	1,806
Leicester	445,600	15	27	31	1·6	8·9	96	158,000	29·8	11,185
Lincoln: Holland	238,393	81	78	26	1·7	3·1	108	—	—	0
Kesteven	404,700	60	66	27	1·8	7·1	81	139,000	30·0	2,098
Lindsey	832,063	60	67	24	1·6	8·2	106	172,000	17·8	11,377
Middlesex	35,503	33	32	9	3·3	7·8	35	17,000	7·8	3,374
Norfolk	975,341	72	75	17	0·4	3·1	114	120,000	9·2	11,090
Northampton	498,385	23	36	32	3·2	12·9	72	129,000	22·1	14,524
Soke of Peterboro'	42,445	55	63	28	0·1	4·4	91	—	—	0
Northumberland	644,735	22	30	8	0·01	1·0	64	—	—	2,415
Nottingham	407,524	45	53	27	0·6	7·8	72	49,000	9·2	2,419
Oxford	386,910	36	51	28	0·06	11·8	49	166,000	34·8	4,096
Rutland	86,451	31	41	17	1·3	7·7	69	33,000	34·1	630
Shropshire	693,232	24	35	19	0·4	2·1	94	—	—	863
Somerset	781,223	16	23	18	1·9	3·9	19	170,000	15·7	347
Stafford	539,382	22	31	22	0·8	2·3	64	—	—	376
E. Suffolk	429,414	66	74	20	5·0	9·9	125	297,000	53·4	49,786
W. Suffolk	285,990	74	81	23	3·3	9·0	94	174,000	44·8	40,376
Surrey	176,516	26	44	17	1·2	9·8	14	175,000	38·2	2,482
E. Sussex	303,894	18	34	23	1·6	9·2	23	53,000	10·5	1,886
W. Sussex	229,769	34	52	26	0·7	7·4	22	149,000	37·2	7,357
Warwick	471,738	21	32	29	1·7	8·7	76	126,000	20·1	11,634
Westmorland	225,322	14	21	0·2	0	0·3	182	—	—	744
Wiltshire	597,455	26	42	28	0·6	13·4	15	86,000	10·1	1,233
Worcester	353,087	29	36	17	3·8	7·6	52	64,000	14·0	3,384
York E.R.	653,617	63	70	22	1·4	5·3	74	30,000	4·1	3,644
York N.R.	813,849	35	42	16	0·44	6·0	63	120,000	8·8	1,338
York W.R.	1,033,752	29	36	24	0·2	5·0	85	—	—	489

attention is paid to the possibilities of mole draining in co-operation with tile draining, as a means of utilising this very cheap method of draining over a much wider range of soils than it has been customary to regard as suitable.

Attention has been called to the progress made in tile draining, to its great expense when done intensively, and to the probability that work of this kind has mainly been in the nature of repairs and renewals rather than comprehensive schemes. The scarcity of labour and of tiles has also had a marked deterrent effect on such activities. Several types of mechanical excavator have been imported from America or evolved in this country for digging trenches to receive tiles and have been operated with success. In the hands of contractors or War Agricultural Executive Committees they obviously provide a partial solution of the labour difficulties.

It is all to the good that the assistance given to the re-conditioning of field ditches should be producing a substantial response. Except in country such as the Fens, it can be assumed that to restore 1 chain of ditch will improve from 1 to 2 acres of land. The state of neglect into which these vital drainage channels have fallen on many farms is appalling. To probe their depths with a walking-stick or a spade, to look at the condition and position of culverts under gateways and roads, and to study their lay-out and dimensions, is to realise the thoroughness of the men who made them originally and the importance which they attached to them. A walk along such a ditch, in winter or in summer, will frequently reveal the evidence of important field outfalls, hidden from view and unable to function properly. It is safe to say that if nothing more were to be done than to re-condition such ditches, it would be found that the tile-drain systems on a substantial acreage of medium and light land would thereby be put into commission once more, would run freely again, and would greatly benefit the land in which they lie. Probably the greatest obstacle to be overcome in this problem at present is the shortage of the necessary labour. Much has been accomplished already by machinery and by the employment of casual and unskilled hands and it ought to be possible greatly to augment the amount of reclamation carried out in this fashion.

APPENDIX

PARTICULARS OF THE SCHEMES OF GOVERNMENT ASSISTANCE FOR DRAINAGE WORK[1]

Grants for farm drainage

Grants towards the cost of ditching and under-drainage are available to owners or occupiers of agricultural land. The rate of grant is 50 per cent of the actual net cost of approved work, subject to the limits set out below. Any farmer whose land needs draining may apply to his County Agricultural Executive Committee for advice as to the most suitable method to employ and for forms of application for grant.

DITCHING

Ditches are the key to the drainage of waterlogged fields. The clearance of neglected ditches may bring into operation existing under-drainage systems and so render new underdrainage unnecessary. For this reason Executive Committees may require ditches to be put into good order before they approve a mole drainage or a tile drainage scheme. The work of clearing the ditches is eligible for grant as a ditching scheme.

Maintenance work, such as brushing and weed-cutting, which should in the ordinary way be carried out every year, cannot be grant-aided, but if ditches have got into a bad state through lack of regular maintenance over a period of years, the work of putting them into proper condition again may be treated as an improvement scheme eligible for grant. The work of cutting back a hedge and clearing trees or stumps, so far as it is necessary to enable ditching work to be done, may be included in a ditching scheme.

If the Executive Committee think it necessary that a ditch which is to be improved under a grant-aided scheme should be fenced, to prevent it being trodden in by stock, they may require a fence to be erected·as a condition of their approval of the scheme.

[1] Reprinted by permission of the Ministry of Agriculture and Fisheries.

A reasonable charge for fencing a ditch improved under a grant-aided scheme may be included as part of the cost of the scheme, but the grant on the fencing will be limited to a certain maximum rate per chain. Fencing should be of a simple type, such as a single strand of barbed wire on wooden posts. Excessive expenditure on fencing will not be grant-aided.

Work on reconstructing and altering culverts and providing drinking bays is eligible for grant provided such work is ancillary to the drainage scheme, but the cost of erecting new culverts on new sites is not eligible for grant.

The work of piping an open watercourse may in special circumstances be approved for grant as a ditching scheme, but before approval is given the Ministry must be satisfied that the work is necessary for improving the drainage of the land and not merely for ease of cultivation.

No maximum (except as regards fencing) is laid down for the cost of improving ditches. Every scheme will be considered on its merits, and applicants must not expect to receive a grant on expenditure which is out of proportion to the probable benefit. Applications for ditching grants should be accompanied by a 6 in. Ordnance Survey sheet or a tracing from such a sheet, indicating the ditches to be improved.

MOLE DRAINAGE

Mole drainage grants are intended to assist in bringing into effective cultivation the particular type of heavy land for which mole drainage is a satisfactory alternative to the more expensive methods of under-drainage. In order to give full and lasting benefit, a mole drainage system may require piped mains, and such work will be treated as an essential part of the scheme.

When the combined length of tiled mains and leads does not exceed one chain per acre drained, grant on the whole scheme will be at the rate of 50 per cent of the actual cost, but may not exceed £2 per acre drained If the combined length exceeds an average of one chain per acre drained, the scheme will be treated as a tile drainage scheme, and grant aided at tile drainage rates.

When a mole drainage scheme is not wholly satisfactory, and it is considered that the cost of a full tile drainage scheme would be too high in relation to the value of the land, the practice of laying a widely spaced network of tiles and drawing moles over them is to be encouraged. A scheme of this kind will, provided the combined length of tiling exceeds an average of one chain per acre drained, be treated as a tile drainage scheme, with the whole of the cost eligible for grant at the approved rate for tile drainage.

The applicant will be required, on completion of the work, to furnish the Executive Committee with an accurate plan certified by a qualified person. The plan should be shown on a 25 in. Ordnance Survey sheet, or a tracing on linen from such sheet, and should indicate the position of the tiled mains and leads, and give detailed measurements.

TILE DRAINAGE

Tile drainage includes all recognised methods of under-drainage other than mole drainage. The piping of an existing watercourse cannot be treated as a tile drainage system.

The grant for tile drainage will be 50 per cent of the actual net cost of approved work, subject to a maximum grant of £15 per acre of benefit. In special cases, the Ministry is prepared to consider a recommendation from a County Agricultural Executive Committee that a higher rate of grant than the normal maximum of £15 per acre should be approved. In such cases, the Ministry will require to be satisfied that (a) efficient drainage cannot be secured by any less expensive form of layout, (b) the unit costs are reasonable in relation to the rest of the county, and (c) the benefits of food production likely to result from the proposed works warrant the expenditure involved. If an applicant is of the opinion that his scheme satisfies these three conditions, he should ask the Committee to consider making a recommendation to the Ministry for a special rate of grant to be approved.

It is a normal condition of tile drainage grant that the applicant shall on completion of the approved work furnish the Executive Committee with an accurate plan on a 25 in. Ordnance Survey sheet, or a tracing from such a sheet, showing the drainage system laid down, together with full measurements, certified by a person who, in the opinion of the Committee, is qualified to give the required certificate. The applicant should consult the Committee as to the information to be given on the plan. The cost of the plan may be included as part of the cost of the scheme. A plan will always be required unless the Ministry is satisfied that the scheme provides only for very small repairs to an existing drainage system, and then only if that system is already shown on a plan which is likely to be permanently available, such as a plan attached to title deeds. Where no plan already exists of a drainage system reconditioned under a grant-aided scheme, a plan will have to be made, but it need show only those parts of the system on which work has been done under the scheme. The plans furnished in connection with schemes will be kept by the Ministry for record, and copies will be obtainable in future at an appropriate charge by persons having an interest in the land concerned.

How to Get the Work Done

If a farmer is unable to find the necessary labour to carry out a drainage scheme, or to arrange with a drainage contractor to undertake it on his behalf, he should ask the Executive Committee whether they can help him. The Committee may be able to carry out the work with their own labour and machinery, if the farmer undertakes to pay the cost (less grant) on completion of the work.

How to Get the Grant

Before any work is begun, applications for grant should be submitted to the Committee in duplicate on the appropriate forms; they should include or be accompanied by an estimate showing clearly how the cost of the scheme has been arrived at. After satisfying themselves that the work is desirable and the estimate reasonable, the Committee will approve the scheme for grant. When approval has been given, *but not before*, work may begin, on the understanding that the grant will not be payable until the Committee certify that the work has been satisfactorily carried out in accordance with the approved scheme.

Any owner or occupier carrying out an approved scheme does so on his own responsibility. The applicant is responsible for observance of the conditions on which the scheme is approved, whether he does the work with his own farm labour or employs a contractor. If the work is supervised by a person not normally employed on the farm or the estate, reasonable fees or wages paid to that person may, at the discretion of the Ministry, be allowed to rank for grant. No allowance can, however, be made for work done by the applicant himself or for time given to supervision by the applicant or one of his regular employees or agents. Any workmen employed on the work must be paid at proper rates not less than the minimum rates payable to unskilled agricultural workers in the district under the appropriate Order of the Agricultural Wages Board. All the workers employed in connection with a scheme must be properly insured under the National Insurance Acts, and applicants and their contractors should also ensure themselves against all risks of injury, damage or loss arising out of the carrying out of the work, whether to their workers or other persons, or to the property of themselves or of any other persons. No payment arising as a result of the failure to effect such insurance will be admitted as part of the cost of the work for the purpose of grant aid.

The date of commencement and the date of completion of any approved works must be notified at once to the Executive Committee. If, while the scheme is in progress, the applicant wishes to vary it in any way he must obtain the Committee's prior approval; otherwise he will run the risk of losing the grant on the whole scheme. When work has been completed, the Committee will make such enquiries as they think necessary, to enable them to certify that the work has been carried out to their satisfaction, in accordance with the approved scheme, and that all the prescribed conditions have been fulfilled. The Committee will require to be supplied with receipted vouchers for the purchase of materials, the hire of machinery or the payment of a contractor. They may also require to be satisfied that the labour costs included in the account have actually been incurred. The Ministry reserves the right to have the work inspected before, during or after completion, by one of the Ministry's Drainage Engineers.

The grant will be paid to the applicant by the Executive Committee, and any enquiries about payment of grant should be addressed to the Agricultural Officer of the Committee and not to the Ministry.

SOME USEFUL DATA

Length

12 inches	= 1 foot
3 feet	= 1 yard
5½ yards	= 1 rod, pole or perch
40 poles	= 1 furlong
8 furlongs	= 1 mile

7·92 inches	= 1 link
100 links	= 1 chain
22 yards	= 1 chain

1 centimetre = 0·3937 inch
1 metre = 39·37 inches
1 inch = 2·54 centimetres

Area

144 square inches	= 1 square foot
9 ,, feet	= 1 ,, yard
30¼ ,, yards	= 1 ,, rod
40 ,, rods	= 1 rood
4 roods	= 1 acre

1 acre = 4840 square yards
1 acre = 10 square chains
1 square mile = 640 acres

1 hectare = 10,000 square metres
= 2·47 acres
1 acre = 0·405 hectare

Volume

2 pints	= 1 quart
4 quarts	= 1 gallon

10 pounds of water = 1 gallon

1,000 cubic centimetres = 1 litre
1 litre = 0·22 gallon
1 gallon = 4·55 litres
1 cubic foot = 6·24 gallons

1 foot × 1 foot × 1 chain = 2·44 cubic yards

Rainfall and run-off

1 inch of rain on 1 acre = 22,635 gallons or 101·1 tons
1 pint per second = 450 gallons per hour
1 cusec = 1 cubic foot per second
= 374 gallons per minute
= 1 inch per acre per hour
1 gallon per acre per hour = 0·003122 litre per hectare per second
1 litre per hectare per second = 320·3 gallons per acre per hour
Drainage co-efficients, i.e. the amounts to be cleared in 24 hours:
1 inch = 943 gallons per acre per hour

Drainage work

Depth: Mole draining, 15 inches with the lightest tackle to 30 inches with the
heaviest. Common practice, 20 to 24 inches
Tile draining in clay land, 18 to 24 inches
in open soils, 30 to 36 inches
See also p. 80.

Distance apart: Mole draining, 3 to 5 yards
Tile draining in clay land, 5 to 8 yards
See also p. 80.

Diameter of pipes: 2½ inches upwards, according to the area served. With a fall of 1 in 200,

a 3 inch pipe will serve 2 to 3 acres
a 4   ,,   ,,   5 to 7   ,,
a 6   ,,   ,,   15 to 20   ,,
a 9   ,,   ,,   45 to 55   ,,
See also p. 131.

Length of channels required to drain one acre:

At intervals of 9 feet,   4,840 feet or 73·3 chains
     ,,   8 yards, 1,815   ,,   27·5   ,,
     ,,   11   ,,   1,320   ,,   20   ,,
     ,,   16   ,,   907   ,,   13·8   ,,
     ,,   22   ,,   660   ,,   10   ,,

A Bibliography of Field Drainage

The Mode of Draining Land, by John Johnstone. Richard Phillips, London, 1801.

Practical Drainage of Land, by H. Hutchinson. Houlston and Stoneman, London, 1844.

Agricultural Drainage, by J. Bailey Denton. Spon, London, 1883.

Land Drainage, by G. S. Mitchell. Land Agents' Record Ltd., London, 1898.

Land Drainage, by C. H. J. Clayton. Country Life and George Newnes Ltd., London, 1919.

Engineering for Land Drainage, by C. G. Elliott. John Wiley and Sons, New York, 1919.

Text-Book of Land Drainage, by J. A. Jeffery. MacMillan Co., New York, 1921.

The Principles of Underdrainage, by R. D. Walker. Chapman and Hall, London, 1929.

Land Drainage, by W. L. Powers and T. H. A. Teeter. John Wiley and Sons, Inc. and Chapman and Hall, London, 1932.

Land Drainage, by B. W. Adkin. Estates Gazette Ltd., London, 1933.

Handbook of Culvert and Drainage Practice. Armco (Pty) Ltd., Lakeside Press, Chicago, 1938.

Land Drainage and Reclamation, by Q. C. Ayres and D. Scoates. McGraw Hill Book Co., New York, 1939.

Drainage and Flood Control, by G. W. Pickles. McGraw Hill Book Co., New York, 1941.

Land Drainage, by R. G. Kendall. Faber and Faber Ltd., London, 1950.

The Draining of Farm Lands, by A. W. Hudson and H. G. Hopewell. Bull. No. 18, Massey Agricultural College, Palmerston North, 1950.

INDEX

Printed in the United States
By Bookmasters